EXCELLENCE IN MATH

MATHEMATICS WORKBOOK

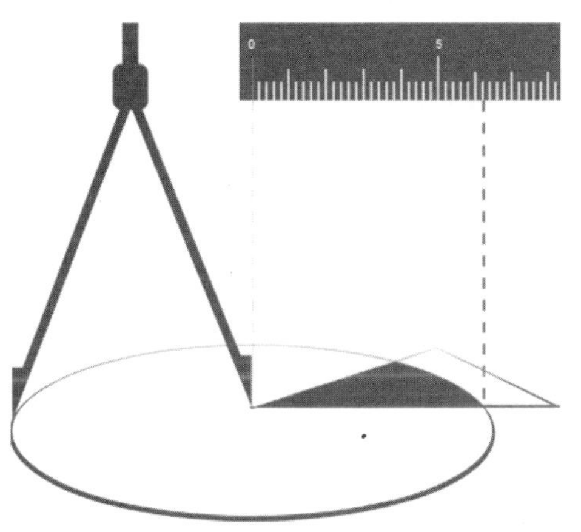

PHOENIX SCIENCE PRESS LTD.

江苏科学技术出版社

Excellence in Math

Mathematics 8 Workbook

Author

Chris Zhang, Ph.D.

Reviewers

Cary Chien, Ph.D. and Griffin Tench

Publishers

Phoenix Science Press Ltd.

Jinglun Culture Media Group

Printed in China

Distributor

Doyen Publishing Inc.

865 W. 70th Avenue, Vancouver

BC V6P 2X6

Tel: 604.721.9698

Fax: 604.323.9698

Email: grace@doyenpublishing.ca

Website: www.doyenpublishing.ca

Excellence in Math

Mathematics 8 Workbook

ISBN 978-7-5537-1843-9

About the Series

Excellence in Math, a series of five workbooks, is designed to specifically cover Grade 8 to Grade 12 secondary school curriculum.

These workbooks were developed through in–depth research of Western Canada (including provinces of British Columbia, Alberta, Manitoba and Saskatchewan) mathematics curriculum and in conjunction with the admission requirements of North American universities and colleges.

Students benefit from using these books to not only complete their high school courses with the aim of attaining high marks, but to also prepare for SAT Math and AP Calculus.

Each workbook has 8 to 10 chapters. Each chapter is clearly structured with the following sections:

Theory and Examples: detailed explanation of main concepts and theories required by the curriculum illustrated by various examples

Practice: plenty of levelled exercises, which aim to train students in both fundamental concepts and advanced problems solving skills

Chapter Test: enable students to review the whole chapter and test their understanding of basic concepts and theories

Full answer keys for practice and chapter tests are provided at the end of each workbook. To help students understand and learn more effectively, solutions are also given to exercises in Level C.

Acknowledgements

We would like to thank all the teachers, students and friends involved in research, writing and editing of this series, especially those at Sager Education. A special thank-you to the following students (in alphabetic order of last name) for their great help in the writing process.

Xiwei Peter Fuqian for Grade 9 and 11

Carvin Mo for Grade 12

Robert Mo for Grade 8

Jerry Wang for Grade 10

Also, we would like to thank our families for their huge help, care and understanding during the period of writing and publishing these books.

CONTENTS

CHAPTER 1 Numbers

1.1 Exponent and Scientific Notation 2
1.2 Percents, Ratios, Proportions and Rates 10
1.3 Squares and Square Roots 17
1.4 Rational Numbers, Irrational Numbers and Number Systems 24
Chapter 1 TEST 29

CHAPTER 2 Percent

2.1 The Percent of a Number 38
2.2 The Percent of Two Numbers 45
2.3 Discount, Tax, GST and PST 50
2.4 Commission Rates and Interest Rates 55
2.5 Per mil, Ppm and Ppb* 61
Chapter 2 TEST 64

CHAPTER 3 Ratios, Proportions, Scales and Rates

3.1 Ratios, Proportions and Scales 68
3.2 Rates and Unit Rates 76
3.3 More Application Problems Regarding Rates and Ratios 83
Chapter 3 TEST 87

CHAPTER 4 Multiplication and Division of Fractions and Integers

4.1 Multiplication and Division of Fractions 92
4.2 Multiplication and Division of Integers 99
4.3 Order of Operations 105
Chapter 4 TEST 111

---

* This section is optional. Teachers and students can decide whether they need to study it.

CHAPTER 5 Variables and Relations

5.1 Variables 117

5.2 Three Ways to Describe Relations between Two Variables 122

5.3 Linear Relations 127

Chapter 5 TEST 134

CHAPTER 6 Linear Equations

6.1 Combining Like Terms and Removing Brackets 139

6.2 Solving Simple Linear Equations 145

6.3 Solving Complicated Equations 152

Chapter 6 TEST 159

CHAPTER 7 Pythagorean Theorem, Combined Shapes and Tessellations

7.1 Pythagorean Theorem 164

7.2 The Perimeter and Area of Combined Shapes 172

7.3 Tessellations 178

Chapter 7 TEST 184

CHAPTER 8 Nets, Surface Area and Volume

8.1 Nets 191

8.2 Surface Area 197

8.3 Volume 204

Chapter 8 TEST 211

CHAPTER 9 3-D Views and Three Views of 3-D Objects

9.1 Three Views from a 3-D View 221

9.2 The 3-D View from Three Views of an Object 225

9.3 3-D Views and Three Views of Combined Objects 230

Chapter 9 TEST 234

CHAPTER 10 Probability

10.1 Probability 239

10.2 Independent Events and Dependent Events 247

Chapter 10 TEST 254

Answers to Practice and Chapter Tests 259

CHAPTER 1

Numbers

1.1 Exponent and Scientific Notation

1.2 Percents, Ratios, Proportions and Rates

1.3 Squares and Square Roots

1.4 Rational Numbers, Irrational Numbers and Number Systems

Chapter 1 TEST

Chapter 1 Numbers

1.1 Exponent and Scientific Notation

Base, Exponent and Power

$2\times2\times2$ can be written as 2^3, which means that 2 is multiplied 3 times. 2 is called the **base**, 3 is called the **exponent** and 2^3 is called a **power**.

Base —— ② ③ —— Exponent

Example 1

Change the following expressions to the power form.

a) $12\times12\times12$ **b)** $2.3\times2.3\times2.3$ **c)** $\frac{9}{16}$ **d)** 81

▶ *Solution*:

a) $12\times12\times12=12^3$

b) $2.3\times2.3\times2.3=(2.3)^3$

c) $\frac{9}{16}=\frac{3}{4}\times\frac{3}{4}=(\frac{3}{4})^2$

d) $81=9^2=3^4$

Exponent Rules

1. $a^1=a$

2. $a^0=1$

3. $a^m\cdot a^n=a^{m+n}$

4. $a^m\div a^n=a^{m-n}$

5. $a^{-m}=\frac{1}{a^m}$

6. $a^{-1}=\frac{1}{a}$

Example 2

Simplify.

a) $7^4\times7^3$ **b)** $3^4\div3^5$ **c)** 8^{-2} **d)** $5^3\div5^{-2}\times5^4$

▶ *Solution*:

a) $7^4\times7^3=7^{4+3}=7^7$

b) $3^4\div3^5=3^{4-5}=3^{-1}=\frac{1}{3}$

c) $8^{-2}=\frac{1}{8^2}=\frac{1}{64}$

d) $5^3\div5^{-2}\times5^4=5^{3-(-2)+4}=5^9$

Example 3

Calculate the following.
$4(3^0+5^1)\div(2^{-1}\times 2^3)$

▶ *Solution*:

$4(3^0+5^1)\div(2^{-1}\times 2^3)=4(1+5)\div 2^{-1+3}=4\times 6\div 2^2=6$

Standard Form and Scientific Notation

Standard Form is the regular way of writing numbers, i. e. 38000, 0.00576.

Scientific Notation is a way of writing a very large number or a very small number. Scientific notation is formed by the product of a number between 1 and 10 but less than 10 and the power of 10. For example, $58\ 000 = 5.8\times 10^4$, and $0.0058=5.8\times 10^{-3}$.

As a general rule, when using scientific notation to express a very large number, the decimal point of the large number moves to the left side until the number becomes a value between 1 and 10, then the number of jumps of the decimal point becomes the positive exponent of the power of 10.

Example 4

Change the following number to scientific notation.
12 900

▶ *Solution*:

$12\ 900=1.29\times 10^4$
When the decimal point of 12 900 (12 900 can be treated as 12 900.0) moves 4 digits to the left, 12 900 becomes 1.29 times of the power of 10 with the exponent of 4.

Example 5

Calculate, using scientific notation.
a) $2.1\times 10^4+1.3\times 10^3$
b) $(2.1\times 10^4)\times(1.3\times 10^3)$
c) $(1.3\times 10^5)\div(2.1\times 10^3)$

▶ *Solution*:

a) $2.1\times 10^4+1.3\times 10^3=21\times 10^3+1.3\times 10^3=(21+1.3)\times 10^3=22.3\times 10^3=2.23\times 10^4$
When adding or subtracting numbers that are written as powers with the equal base, the exponents have to be the same. Always change the exponents first before these operations.

b) $(2.1\times10^{4})\times(1.3\times10^{3})=2.1\times1.3\times10^{3+4}=2.73\times10^{7}$

c) $(1.3\times10^{5})\div(2.1\times10^{3})=(1.3\div2.1)\times10^{5-3}=0.619\times10^{2}=6.19\times10^{1}$

When using scientific notation to express a very small number, the decimal point of the small number moves to the right side until the number becomes a value between 1 and 10 but less than 10, then the number of jumps of the decimal point becomes the negative exponent of the power of 10.

Example 6

Change the following number to scientific notation.

0.0098

▶ *Solution*:

$0.0098=9.8\times10^{-3}$

The decimal point of 0.0098 moves 3 digits to the right, so 0.0098 becomes 9.8 times the power of 10 with exponent -3.

Example 7

Calculate, using scientific notation.

a) $2.7\times10^{-2}-4.5\times10^{-3}$

b) $(2.7\times10^{-2})\times(4.5\times10^{-3})$

▶ *Solution*:

a) $2.7\times10^{-2}-4.5\times10^{-3}=27\times10^{-3}-4.5\times10^{-3}=(27-4.5)\times10^{-3}=22.5\times10^{-3}=2.25\times10^{-2}$

b) $(2.7\times10^{-2})\times(4.5\times10^{-3})=2.7\times4.5\times10^{(-2)+(-3)}=12.15\times10^{-5}=1.215\times10^{-4}$

Scientific Notation on a Calculator

When using a scientific calculator, we sometimes see results expressed as 1.15E10 or 6.23E−8.

As we know, the general form can be written as:

$$a\times10^{b}$$

where $1\leqslant|a|<10$ and b is an integer.

When scientific notation is displayed on a calculator, the "×10" part is usually replaced by either "E" or "e". For example, 3.65×10^{2} is expressed as "3.65E2", and 7.59×10^{-5} may be expressed as "7.59e−5".

Practice

■Level A

1. Complete the table.

	Power	Base	Exponent	Standard form
a)	3^2			
b)	2^4			
c)	4^3			
d)	5^2			
e)	8^3			
f)	9^2			

2. Change the following expressions to power form.

a) 5×5 **b)** 9×9 **c)** 10×10

d) 13×13 **e)** 121×121 **f)** 32×32

3. Change to exponential form and calculate.

a) 102×102 **b)** 0.5×0.5 **c)** 1.2×1.2

4. Write as a power with base of 3.

a) 9 **b)** 81 **c)** 27

5. Simplify and change the following numbers to exponential form.

a) $2^2\times2^3$ **b)** 3×27 **c)** $3^3\times3^5$

6. The area of a square is 256 m^2. What is the length of one side?

7. The perimeter of a rectangle is 90 cm. If the width is 20 cm, what is the area of the rectangle?

8. Change the following numbers to scientific notation.

a) 23 000 **b)** 3 000 000 000 **c)** 1 000 000

d) 100 000 000 **e)** 3 200 000 **f)** 570 000

9. Change the following numbers to standard form.

a) 1×10^3 **b)** 2×10^3 **c)** 5×10^4

d) 3.33×10^1 **e)** 2.5×10^0 **f)** 5.2×10^4

10. Change the following numbers to scientific notation.

a) 0.02 **b)** 0.05 **c)** 0.0003

d) 0.0005786 **e)** 0.0000602 **f)** 0.0005003

11. Change the following numbers to standard form.

a) 2.1×10^{3} b) 3.12×10^{-2} c) 1.2×10^{-5}

d) 1.7285×10^{-5} e) 2.519×10^{-6} f) 5.101×10^{-6}

■Level B

1. Write the following numbers in exponential form and calculate.

a) $\frac{4}{5}\times\frac{4}{5}\times\frac{4}{5}$ b) $0.7\times0.7\times0.7$ c) $13\times13\times13$

2. Simplify.

a) $3\times3^{2}\times3^{3}$ b) $2^{5}\times2^{10}\times2^{2}$ c) $9\times9^{2}\times9^{3}$

3. Complete the following patterns. If possible, change them to exponential form.

a) 1, 2, 4, 8, ____, ____, ____ b) 2, 5, 8, 11, ____, ____, ____

c) 1, 5, 25, 125, ____, ____, ____ d) 2, 6, 18, 54, ____, ____, ____

4. Write as a product of powers with bases of 3 or 7.

a) 63 b) 147 c) 441

5. Calculate.

a) $3.1\times10^{2}+2.1\times10^{2}$

b) $4.5\times10^{3}+5.2\times10^{3}$

c) $-5.0\times10^{3}+3.2\times10^{3}$

d) $9.7\times10^{4}-4.9\times10^{4}$

6. Calculate.

a) $2.78\times10^{-2}+3.57\times10^{-2}$

b) $3.26\times10^{-3}+2.87\times10^{-3}$

c) $9.8\times10^{-3}-2.11\times10^{-3}$

d) $3.97\times10^{-4}-1.98\times10^{-4}$

■Level C

1. Calculate the following powers in standard form.

a) $(-5)^{3}$

b) $(-3)^{4}$

c) $(-2)^{6}$

2. Express the following numbers in standard form.

a) 10^{0}

b) 10^{-1}

c) -2^{-2}

3. Write as a product of powers with bases of 5, 7 and/or 9.

a) 175 **b**) 2025 **c**) 2401

4. Change the following numbers to scientific notation.

a) $8.72\times10^{5}\times2.1\times10^{-3}$ **b**) $1.27\times10^{-2}\times2.56\times10^{-5}$

c) $7680\times2.9\times10^{-3}$ **d**) $3.56\times10^{-3}\times679\,000$

5. Change to scientific notation, then calculate.

a) $(-2.25)\times57\,600\times10^{-3}$ **b**) $3.63\times10^{5}\times2.3\times10^{-5}$

c) $10\,080\div10^{3}\times2400$ **d**) $3.5\times10^{4}\div(7.0\times10^{-2})$

1.2 Percents, Ratios, Proportions and Rates

Percents (Percentage)

A **percentage** is a ratio expressed as a number out of 100. When given a ratio, the corresponding percentage is calculated by dividing the terms of the ratio and multiplying the answer by 100.

Example 1

Change the fractions to percents.

a) $\frac{7}{100}$ **b)** $\frac{23}{100}$ **c)** $\frac{1}{2}$ **d)** $\frac{27}{40}$

▶ *Solution*:

a) $\frac{7}{100}=7\%$

b) $\frac{23}{100}=23\%$

c) $\frac{1}{2}=\frac{1\times50}{2\times50}=\frac{50}{100}=50\%$

d) $\frac{27}{40}=\frac{27\times2.5}{40\times2.5}=\frac{67.5}{100}=67.5\%$

Example 2

9 small equal triangles combine to form a big equilateral triangle. What percent of the big triangle is shaded?

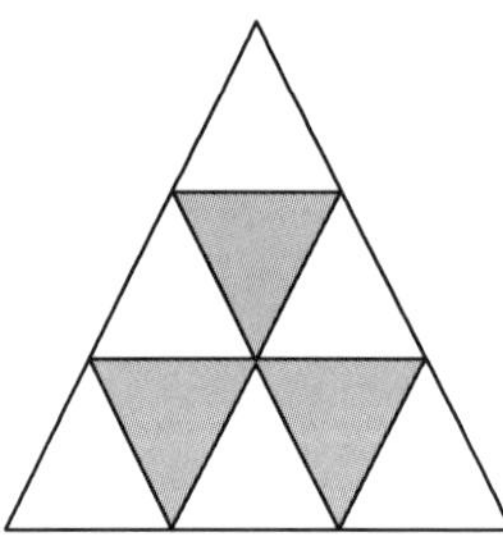

▶ *Solution*:

$3\div9=0.333\cdots\times100\%=33.\overline{3}\%$

The shaded parts take up $33.\overline{3}\%$ of the big triangle.

Ratios

A **ratio** shows the relative sizes of two or more values. A ratio can be written in fraction form, for example, $\frac{2}{3}$. This ratio can also be written as the form 2∶3 or 2 to 3.

Like fractions, a ratio can be written as an equivalent ratio by multiplying the terms of the ratio by the same factor. For example, $\frac{2}{3}=\frac{2}{3}\times\frac{2}{2}=\frac{4}{6}$, the denominator and the numerator are both multiplied by 2.

Example 3

Change the following ratios to the lowest forms.

$\frac{45}{90}$, 30 : 14, 11 to 66

▶ *Solution*:

$\frac{45}{90}=\frac{1}{2}$, 30 : 14=15 : 7, 11 to 66=1 to 6

Proportions

An equation that states that two ratios are equal is called a **proportion.** When a proportion is missing a term, the missing term can be found by cross multiplication.

Example 4

Solve for x in the following proportion: $\frac{x}{15}=\frac{5}{30}$.

▶ *Solution*:

The products of the pairs of cross terms are equal.

$\frac{x}{15}\times\frac{5}{30}$, $30x=5\times15$, $30x=75$, $x=\frac{75}{30}=2.5$

Example 5

Solve for x in the following proportion: $8 : x=4 : 32$.

▶ *Solution*:

The inside product is equal to the outside product (the product of the two inside terms is equal to the product of the two outside terms).

8 : x = 4 : 32 (Inside: x and 4; Outside: 8 and 32), $4x=8\times32$, $4x=256$, $x=\frac{256}{4}=64$

A proportion can include more than two ratios, and can have more than one missing numbers.

Example 6

Solve for x and y.

$\frac{3}{4}=\frac{x}{2}=\frac{6}{y}$

▶ *Solution*:

Separate the two expressions.

(1) $\frac{3}{4}=\frac{x}{2}$, $4x=2\times3=6$, $x=\frac{6}{4}=1.5$

(2) $\frac{3}{4}=\frac{6}{y}$, $3y=6\times4=24$, $y=\frac{24}{3}=8$

Rates

A **rate** is a comparison of two quantities with different units. For example, a store has a sale where you can get 2 shirts for \$50. This can be expressed as $\frac{50 \text{ dollars}}{2 \text{ units}}$.

Unit Rates

A **Unit Rate** compares two quantities, with a second quantity of one unit. For example: the price of apples is \$2/kg(\$2 one kilogram). The speed of the car is 80 km/hr(80 km one hour).

Example 7

Which price is better for buyers, 20 kg for \$25.00 or 3 kg for \$4.50?

▶ *Solution*:

$\frac{\$25.00}{20\text{ kg}}=\frac{\$1.25}{\text{kg}}=\$1.25/\text{kg}$

$\frac{\$4.50}{3\text{ kg}}=\frac{\$1.50}{\text{kg}}=\$1.50/\text{kg}$

The first one is better because \$1.25/kg is cheaper than \$1.50/kg.

Practice

■ Level A

1. What percent does the shaded part take up in each figure?

a)

b)

c)

2. Draw a picture with the following requirements.

a) a square that is 30% shaded in 2 different ways

b) a circle that is 25% shaded in 2 different ways

3. Change the decimals to percents.

a) 0. 5 **b**) 0. 7 **c**) 0. 9

d) 0. 98 **e**) 0. 25 **f**) 0. 50

4. Simplify the ratios to lowest terms.

a) $\frac{4}{16}$ **b**) $\frac{7}{49}$ **c**) $\frac{11}{121}$

d) 3 : 27 **e**) 15 : 225 **f**) 12 : 60

5. Solve for the missing term.

a) $2:3=x:27$ **b**) $3:x=5:35$ **c**) $12:24=4:x$

d) $\frac{15}{10}=\frac{x}{150}$ **e**) $\frac{x}{7}=\frac{7}{9}$ **f**) $\frac{4}{3}=\frac{x}{9}$

6. Jack runs 2. 0 km in 30 minutes. At the same rate, how far can he run in 70 minutes? How long does it take him to run 3. 5 km?

Level B

1. Change the fractions to percents.

a) $\frac{1}{2}$ **b)** $\frac{1}{4}$ **c)** $\frac{1}{5}$

d) $\frac{3}{5}$ **e)** $\frac{3}{4}$ **f)** $\frac{2}{5}$

2. Change the percents to decimals.

a) 23% **b)** 34% **c)** 56%

d) 8% **e)** 5% **f)** 9%

3. The price of a fax machine drops from \$90 to \$60. By what percent does the price decrease?

4. Solve for x and round to the nearest tenth.

a) $\frac{31}{27}=\frac{x}{50}$ **b)** $\frac{x}{16}=\frac{49}{35}$ **c)** $\frac{12}{x}=\frac{36}{22}$

d) $12:x=61:27$ **e)** $x:25=30:90$ **f)** $15:35=5:x$

5. Solve for x.

a) $2:3:4=4:x:y$

b) $4:3:2=x:15:y$

c) $10:x:50=5:30:y$

d) $x:y:20=5:10:45$

6. The sale price of a math textbook at 30% discount is \$27. What was the original price?

■Level C

1. Change the numbers to percents and round to the nearest hundredth.

a) 23.3

b) $1\frac{7}{9}$

c) 2.555…

d) $5\frac{1}{3}$

e) $2\frac{2}{99}$

f) 3.01001001…

g) 0.0092

h) 0.9292…

i) $\frac{92}{99}$

j) $\frac{27}{97}$

k) $\frac{13}{15}$

l) $1\frac{9}{99}$

2. Tom gets \$1000 per week after he received a 25% pay increase. What was his salary before the increase?

3. Select the better deal.

a) 4 kg of bananas for \$2.86 or 11 kg for \$7.81?

b) 18 hours of homework in 13 days or 21 hours of homework in 15 days? (Assume students like to do less homework each day.)

4. Solve for x of each proportion and round to the nearest tenth.

a) $x : 3 : 5 = 17 : y : 35$

b) $x : \frac{1}{2} : \frac{3}{5} = 0.75 : y : 3.5$

c) $19 : x : 77 = 38 : 55 : y$

d) $\frac{2}{3} : 0.8 : y = x : 0.2 : 1.6$

5. The speed of sound is 1460 m/s in water. How far does sound travel in 4.5 seconds in water? How long does sound take to cross 438 km in the sea?

1.3 Squares and Square Roots

Squares

If two of the same numbers are multiplied, the product is called the number's **square.** For instance, $3\times3=3^2$, 3^2 is called three squared.

Perfect Squares

A **perfect square** is an integer that is the square of an integer. In other words, it is the product of an integer multiplied with itself. For example, 16 is a perfect square, since it can be written as 4×4. The perfect square table is very useful.

Perfect Square Table

Numbers	1	2	3	4	5	6	7	8	9	10	11	12	13	14	15	16	17	18	19	20
Perfect Square	1	4	9	16	25	36	49	64	81	100	121	144	169	196	225	256	289	324	361	400

Square Roots

$3^2=9$, and 3 is the number whose square is 9, so 3 is called the **square root** of 9. The form $\sqrt{9}$ is called the radical form, and the $\sqrt{}$ symbol is called the radical sign.

Example 1

Find the following square roots.

$\sqrt{81}$

▶ *Solution*:

$\sqrt{81}=\sqrt{9^2}=9$

So $\sqrt{\text{Perfect Square}}=$number of the base of the square root.

A perfect square is a number whose square root is a whole number. The square root is the opposite operation of a square, and vice versa.

Square Root Properties

$\sqrt{a^2}=a$, $(\sqrt{a})^2=a$, $\sqrt{ab}=\sqrt{a}\times\sqrt{b}$, $\sqrt{\frac{a}{b}}=\frac{\sqrt{a}}{\sqrt{b}}$, where a and b are both greater than zero.

Example 2

Find the square roots of these numbers.

a) $\sqrt{49}$ **b)** $\sqrt{441}$ **c)** $\sqrt{\frac{121}{225}}$

▶ *Solution*:

a) $\sqrt{49}=\sqrt{7^2}=7$

b) $\sqrt{441}=\sqrt{9\times 49}=\sqrt{9}\times\sqrt{49}=3\times 7=21$

c) $\sqrt{\frac{121}{225}}=\sqrt{\frac{11^2}{15^2}}=\frac{\sqrt{11^2}}{\sqrt{15^2}}=\frac{11}{15}$

Estimating Non-perfect Square Roots

Example 3

Estimate the value of $\sqrt{6}$ without a calculator and round to the nearest tenth and the nearest hundredth.

▶ *Solution*:

First we must find two consecutive integers (a and b) between which $\sqrt{6}$ lies. We find the two closest perfect square numbers that are less than 6 and greater than 6, respectively.

As we know $4<6<9$, that is $2^2<6<3^2$

So $2<\sqrt{6}<3$

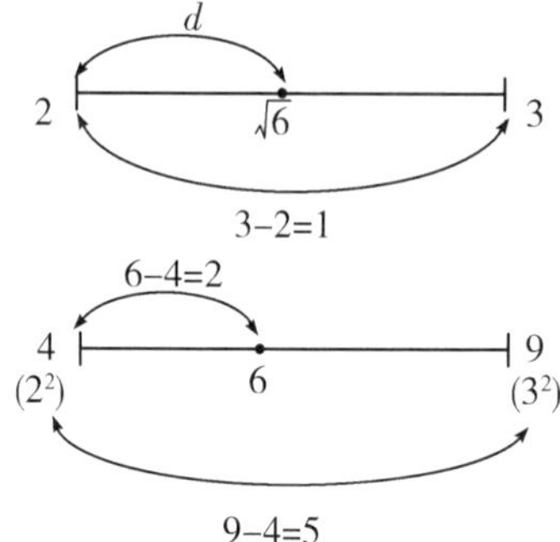

Let $\sqrt{6}=2+d$, we can obtain d by assuming the proportion:

$\frac{d}{3-2}=\frac{6-4}{9-4}\longrightarrow\frac{d}{1}=\frac{2}{5}$

$d=0.4$

So $\sqrt{6}\doteq 2.4$ (round to the nearest tenth)

Similarly, as we know

$5.76<6<6.25$, that is $2.4^2<6<2.5^2$

So $2.4<\sqrt{6}<2.5$

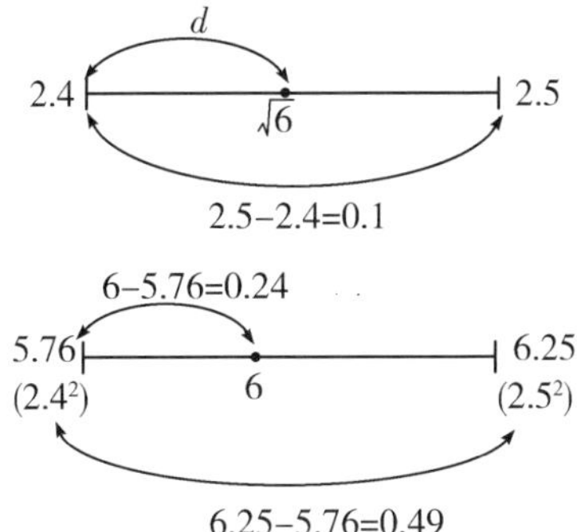

Let $\sqrt{6}=2.4+d$, we can obtain d by assuming the proportion:

$$\frac{d}{2.5-2.4}=\frac{6-5.76}{6.25-5.76} \longrightarrow \frac{d}{0.1}=\frac{0.24}{0.49}$$

$d=0.05$

So $\sqrt{6}\doteq 2.4+0.05\doteq 2.45$ (round to the nearest hundredth)

Example 4

Estimate $\sqrt{600}$, $\sqrt{60}$, $\sqrt{0.6}$ and $\sqrt{0.06}$ to the nearest tenth.

▶ *Solution*:

$\sqrt{600}=\sqrt{6\times 100}=\sqrt{6}\times\sqrt{100}=\sqrt{6}\times 10\doteq 2.45\times 10\doteq 24.5$ (using the estimated value from Example 3)

For $\sqrt{60}$, since $49<60<64$, that is $7^2<60<8^2$.

So $\sqrt{60}\doteq 7+\frac{60-49}{64-49}=7+\frac{11}{15}\doteq 7+0.7\doteq 7.7$

$$\sqrt{0.6}=\sqrt{\frac{60}{100}}=\frac{\sqrt{60}}{\sqrt{100}}=\frac{\sqrt{60}}{10}\doteq\frac{0.77}{10}\doteq 0.8$$

$$\sqrt{0.06}=\sqrt{\frac{6}{100}}=\frac{\sqrt{6}}{\sqrt{100}}=\frac{\sqrt{6}}{10}\doteq\frac{2.4}{10}\doteq 0.2$$

Practice

■Level A

1. Write the squares of every integer from 1 to 20.

2. Write the following perfect squares as powers.

a) 36 **b)** 49 **c)** 64

d) 81 **e)** 144 **f)** 225

g) 121 **h)** 256 **i)** 289

j) 196 **k)** 324 **l)** 361

3. Write the following numbers in exponential form.

a) 0.01 **b)** 0.0009 **c)** 0.000144

d) 1.96 **e)** 0.0081 **f)** 0.0289

g) $\frac{36}{49}$ **h)** $\frac{121}{225}$ **i)** $\frac{144}{324}$

j) $\frac{256}{196}$ **k)** $\frac{225}{400}$ **l)** $\frac{196}{361}$

4. Find the square roots without using a calculator.

a) $\sqrt{64}$ **b)** $\sqrt{144}$ **c)** $\sqrt{361}$

d) $\sqrt{40\ 000}$ **e)** $\sqrt{16\ 900}$ **f)** $\sqrt{25\ 600}$

g) $\sqrt{0.36}$ **h)** $\sqrt{2.25}$ **i)** $\sqrt{2.89}$

j) $\sqrt{\frac{1}{4}}$ k) $\sqrt{\frac{169}{2500}}$ l) $\sqrt{\frac{289}{324}}$

5. A rectangular prism with a square base has a length of 5 cm and a volume of 175 cm^3. What are the dimensions of the prism?

Level B

1. Write each number as a power or as a product of powers.

a) 72 **b)** 52 **c)** 80

d) 243 **e)** 108 **f)** 100

g) 1331 **h)** 90 **i)** 666

j) 432 **k)** 88 **l)** 143

2. Estimate the following square roots to the nearest tenth without using a calculator, then check your answers with a calculator and round to the nearest thousandth.

a) $\sqrt{5}$ **b)** $\sqrt{24}$ **c)** $\sqrt{51}$

d) $\sqrt{70}$ **e)** $\sqrt{132}$ **f)** $\sqrt{156}$

g) $\sqrt{0.06}$ h) $\sqrt{99}$ i) $\sqrt{54}$

j) $\sqrt{110}$ k) $\sqrt{67}$ l) $\sqrt{11}$

3. Bob is twice as old as Tim, and the sum of their ages is 27. How old is each boy?

■Level C

1. Calculate the following powers.

a) $(-5)^3$ b) $(-4)^4$ c) -5^4

d) 5^{-2} e) $(-10)^3$ f) -10^{-2}

g) $(-0.3)^2$ h) $(-0.3)^{-2}$ i) -0.3^2

j) $-(-3)^2$ k) $-(-3)^{-2}$ l) -3^{-2}

2. Find the square roots if possible without using a calculator.

a) $\sqrt{-144}$ b) $\sqrt{-144^{-1}}$ c) $\sqrt{(-4)^2}$

d) $\sqrt{-4^2}$ **e)** $\sqrt{-4^{-2}}$ **f)** $\sqrt{13^{-2}}$

g) $\sqrt{-3^2}$ **h)** $\sqrt{-3^{-2}}$ **i)** $\sqrt{\frac{(-2)^{-2}}{(-4)^{-2}}}$

j) $(\sqrt{-0.25})^2$ **k)** $\sqrt{(-0.25)^2}$ **l)** $\sqrt{-0.25^2}$

3. Estimate the square roots and round to the nearest hundredth.

a) $\sqrt{3}$ **b)** $\sqrt{5}$ **c)** $\sqrt{2}$

d) $\sqrt{13}$ **e)** $\sqrt{7}$ **f)** $\sqrt{6}$

g) $\sqrt{27}$ **h)** $\sqrt{8}$ **i)** $\sqrt{10}$

1.4 Rational Numbers, Irrational Numbers and Number Systems

Rational Numbers

A number which can be changed to a simple fraction (a fraction whose numerator and denominator are integers) is called a **Rational Number.** Rational numbers include natural numbers, whole numbers, integers, simple fractions, and repeating and terminating decimals.

Rational numbers can be changed to a simple fraction.

$0.3=\frac{3}{10}$, $0.333\cdots=\frac{1}{3}$, $0.\overline{13}=\frac{13}{99}$, $17\%=\frac{17}{100}$, $\sqrt{\frac{4}{9}}=\frac{2}{3}$

Irrational Numbers

A number which cannot be changed to a simple fraction is an **Irrational Number.**

π, $\sqrt{3}$, $2\sqrt{3}$, e, 0.121221222…

Comparing and Ordering Real Numbers

Real numbers include rational numbers and irrational numbers. When comparing or ordering real numbers, all numbers should be changed to the same form, such as decimals.

Example 1

Order the following numbers from smallest to largest.

π, $\sqrt{6}$, -5.9, 0, $\frac{7}{9}$, $1.\overline{2}$, 59%

▶ *Solution*:

Change all the numbers to decimals:

$\pi\approx3.14$, $\sqrt{6}\approx2.45$, $\frac{7}{9}=0.777\cdots\approx0.78$, and $59\%=0.59$, so the order is:

-5.9, 0, 59%, $\frac{7}{9}$, $1.\overline{2}$, $\sqrt{6}$, π

The Diagram of Number Systems

The diagram of **Number Systems** includes all types of numbers learned so far.

Diagram of Number Systems

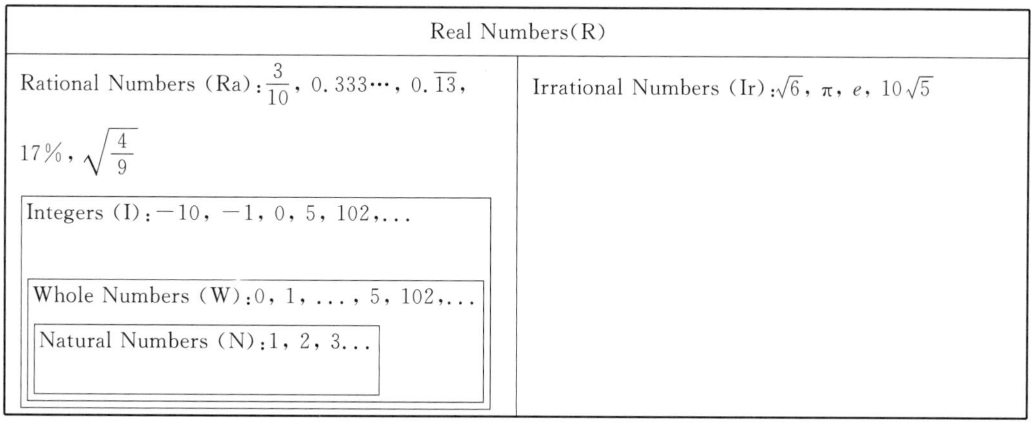

Real Numbers(R)	
Rational Numbers (Ra): $\frac{3}{10}$, 0.333…, $0.\overline{13}$, 17%, $\sqrt{\frac{4}{9}}$ Integers (I): −10, −1, 0, 5, 102,... Whole Numbers (W): 0, 1, ..., 5, 102,... Natural Numbers (N): 1, 2, 3...	Irrational Numbers (Ir): $\sqrt{6}$, π, e, $10\sqrt{5}$

Example 2

Which one is bigger?

a) $(-2)^4$ or -2^4 **b)** 3^{-2} or -3^2 **c)** 73% or 73 **d)** 5^{-2} or $5^{\frac{1}{2}}$

▶ *Solution*:

a) $(-2)^4=(-2)\times(-2)\times(-2)\times(-2)=16$

$-2^4=-2\times2\times2\times2=-16$

So $(-2)^4>-2^4$

b) $3^{-2}=\frac{1}{3^2}=\frac{1}{9}$, $-3^2=-9$

So $3^{-2}>-3^2$

c) 73%=0.73, so 73%< 73

d) $5^{-2}=\frac{1}{5^2}=\frac{1}{25}=0.04$, $5^{\frac{1}{2}}=\sqrt{5}$

So $5^{-2}<5^{\frac{1}{2}}$

Practice

■Level A

1. Draw a diagram explaining number systems.

2. Put a check mark (√) in the space to define the set or sets each number belongs to.

		Set of Numbers					
	Number	N	W	I	Ra	Ir	R
a)	-4						
b)	0						
c)	9						
d)	$0.\overline{37}$						
e)	0.37						
f)	$\sqrt{3}$						
g)	$4\sqrt{3}$						
h)	π						
i)	$\sqrt{\frac{9}{16}}$						
j)	-0.98						

3. Show the following numbers on the number line.

$-2\frac{1}{3}$, -1.3, $-\sqrt{2}$, $-\sqrt{9}$, 2^2, $(-1)^3$, -2^2, 3^{-1}, $0.\overline{3}$

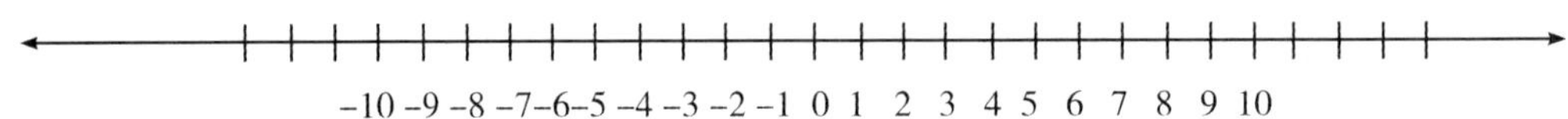

■Level B

1. Express as fractions.

a) 0.17 **b)** -1.005 **c)** 0.025

d) $2:3$ **e)** 3 to 5 **f)** 15 to 3

g) 0.47 **h)** 0.03 **i)** 0.9

j) 1.07 **k**) 2.13 **l**) 3.172

2. Compare each pair of numbers.

a) 0.5 or $0.\overline{5}$ **b**) 0.9 or $0.\overline{9}$ **c**) 1 or $0.\overline{9}$

d) -0.3 or $0.\overline{3}$ **e**) -0.3^0 or $(-3)^0$ **f**) $\frac{1}{3}$ or $0.\overline{3}$

g) 0.8^{-1} or $0.\overline{8}$ **h**) -0.8^{-1} or -0.8^1 **i**) $-0.\overline{08}$ or $-\frac{4}{50}$

j) -5^2 or $(-5)^2$ **k**) -5^2 or $(-5)^{-2}$ **l**) -5^{-2} or $(-2)^{-2}$

■Level C

1. Change the following numbers to fractions if possible.

a) $0.\overline{135}$ **b**) 0.135 **c**) $0.1\overline{35}$

d) $1.1\overline{6}$ **e**) 1.16 **f**) $1.122122122\cdots$

g) $3.13133133313333\cdots$ **h**) 3.1313133 **i**) $3.13131\overline{3}$

j) $\sqrt{\frac{4}{16}}$ k) $\sqrt{-\frac{25}{36}}$ l) $\sqrt[3]{-\frac{8}{27}}$

2. Estimate the value of the following irrational numbers and round to the nearest hundredth.

a) $\sqrt{7}$ b) $\sqrt{2}$ c) $\sqrt{14}$

d) $\sqrt{21}$ e) $\sqrt{3}$ f) $\sqrt{5}$

g) $\sqrt{101}$ h) $\sqrt{1.01}$ i) $\sqrt{10.1}$

j) $\sqrt{231}$ k) $\sqrt{2.31}$ l) $\sqrt{23.1}$

Chapter 1 TEST

1. Complete the table.

	Power	Base	Exponent	Standard form
a)	6^2			
b)	$(-2)^4$			
c)	1^8			
d)	$(-5)^2$			
e)	0.4^3			
f)	$(-11)^2$			

2. Express the following as powers.

a) $12\times12\times12$ **b)** 11×11 **c)** $7\times7\times7\times7$

d) $21\times21\times21\times21\times21$ **e)** 202×202 **f)** $15\times15\times15$

3. Express as powers.

a) 5.5×5.5 **b)** $\frac{2}{3}\times\frac{2}{3}$ **c)** $\frac{3}{4}\times\frac{3}{4}$

4. Write as a power with base of 3.

a) 3 **b)** 243 **c)** 729

5. Simplify by writing the following numbers as a single power.

a) $4^2\times2$ **b)** $3^2\times27$ **c)** $2^{100}\times2^{201}$

6. Change the following numbers to scientific notation.

a) 56 789 000 **b)** 100 100 000 **c)** 120 210 000

7. Change from scientific notation to standard form.

a) -2.56×10^{4} **b)** -2.7×10^{6} **c)** 5.36×10^{5}

d) $1.78\times10^{3}+2.1\times10^{4}$ **e)** $3.31\times10^{4}-1.25\times10^{3}$

8. Change the following numbers to scientific notation.

a) $\frac{1}{28}$ **b)** $\frac{3}{35}$ **c)** $\frac{3}{99}$

d) -0.000987 **e)** 0.00703 **f)** -0.0010001

9. Change from scientific notation to standard form.

a) -2.51×10^{-3} **b)** -3.7×10^{-2} **c)** -4.281×10^{-3}

d) $5.45\times10^{-1}-2.42\times10^{-1}$ **e)** $4.47\times10^{-2}+5.54\times10^{-2}$

10. Write the following numbers as powers and calculate.

a) $\frac{1}{2}\times\frac{1}{2}\times\frac{1}{2}\times\frac{1}{2}$ **b)** $0.3\times0.3\times0.3\times0.3$ **c)** $\frac{3}{4}\times\frac{3}{4}\times\frac{3}{4}$

11. Simplify.

a) $3^8 \div 3^4 \times 3^4$ **b)** $2^{10} \times 2^{15} \div 2^9$ **c)** $9 \times 9^4 \times 9^5 \div 9^9$

12. Write as powers with bases of 2, 3 or 7 or as products of powers.

a) 196 **b)** 343 **c)** 81 **d)** 729

13. Calculate and write your answers in scientific notation.

a) $4.5 \times 10^5 \times 2.5 \times 10^2$ **b)** $2.7 \times 10^3 \times 3.1 \times 10^3$

c) $8.8 \times 10^2 \div (2.0 \times 10^2)$ **d)** $9.8 \times 10^3 \div (4.9 \times 10^3)$

14. Calculate and write your answers in scientific notation.

a) $9.8 \times 10^{-3} - 2.11 \times 10^{-3}$ **b)** $3.97 \times 10^{-4} - 1.98 \times 10^{-4}$

15. Calculate the following powers.

a) -5^4 **b)** $(-2)^6$ **c)** $(-8)^3$

16. Express in standard form and calculate.

a) 10×2^3 **b)** $(-2)^{-2}$ **c)** -3^{-2}

17. Write as powers with bases of 5, 7 or 9, or as products of powers.

a) 11 025 **b)** 1225 **c)** 1575

18. Change the following numbers to scientific notation.

a) $3.25\times 10^{-3}+4.35\times 10^{-2}$ **b)** $9.0\times 10^{3}-9.0\times 10^{2}$

c) $8.98\times 10^{5}-21.37\times 10^{3}$ **d)** $1.237\times 10^{-2}-2.37\times 10^{-4}$

19. Calculate and change to scientific notation.

a) $(-9800)\times 10^{-5}\div 100$ **b)** $240\ 000\times 10^{-4}\div(6\times 10^{-2})$

c) $0.00033\div(-0.0011)\div 10^{-5}$ **d)** $0.327\times 1.5\div(-10)^{3}\times(-2)^{4}$

20. What percent does the shaded part take up in each figure?

a) 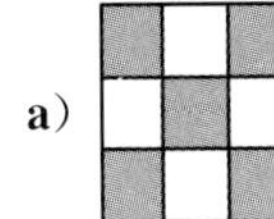

b) 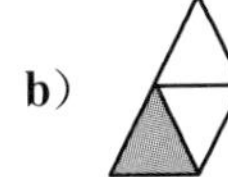

c) 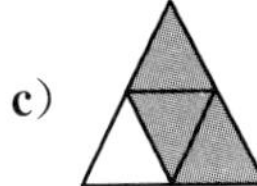

21. Draw a picture with the following requirements.

a) a rectangle which is 20% shaded in 2 different ways

b) a triangle which is half-shaded in 2 different ways

22. Change the decimals to percents.

a) 1.15 **b)** 2.35 **c)** 6.77

d) 0.05 **e)** 10.1 **f)** 8.09

23. Simplify the ratios to lowest terms.

a) 18 to 3 **b)** 11 : 132 **c)** 25 : 700

d) 0.8 : 0.16 **e)** 0.3 : 2.7 **f)** 16 : 0.8

24. Solve for the missing terms.

a) $\frac{20}{x}=\frac{15}{90}$ **b)** $\frac{13}{x}=\frac{13}{5}$ **c)** $\frac{5}{13}=\frac{x}{26}$

d) $2:0.5=100:x$ **e)** $x:1.2=1.2:2.4$ **f)** $x:7=14:1$

25. Change the fractions to percents.

a) $\frac{97}{100}$ **b)** $\frac{77}{100}$ **c)** $\frac{50}{50}$

d) $\frac{13}{20}$ **e)** $\frac{21}{25}$ **f)** $\frac{49}{50}$

26. Change the percents to decimals.

a) 116% **b)** 220% **c)** 100%

d) 0.5% **e)** 0.25% **f)** 10.01%

27. Solve for x and/or y and round to the nearest tenth.

a) $\frac{0.8}{2.4}=\frac{x}{5.6}$ **b)** $\frac{x}{3.2}=\frac{12.8}{16}$ **c)** $\frac{5.4}{0.27}=\frac{3.7}{x}$

d) $\frac{3}{2}:4.5=\frac{4}{5}:x$ **e)** $1.8:\frac{3}{4}=x:0.5$

28. Solve for x.

a) $x:y:100=10:50:300$ **b)** $12:y:x=4:8:96$

c) $\frac{x}{8}=\frac{56}{y}=\frac{72}{16}$ **d)** $\frac{22}{242}=\frac{y}{11}=\frac{33}{x}$

29. Select the better deal.

a) 12 books for \$20.00 or 11 books for \$18.70

b) \$398 for 3 nights in a hotel or \$625 for 5 nights in a hotel

30. Solve for x or y and round to the nearest tenth.

a) $2x:98=6:10$ **b)** $24:3y=9:27$

c) $2y : 3x : 100 = 50 : 90 : 200$

d) $20 : 3x : 30 = 3y : 20 : 10$

31. Put a check mark (√) in the space to define the set or sets each number belongs to.

		Set of Numbers					
	Number	N	W	I	Ra	Ir	R
a)	-18						
b)	0						
c)	$0.\overline{29}$						
d)	$\sqrt{7}$						
e)	π						
f)	$\sqrt{\frac{4}{25}}$						

32. Change to the fractional form and simplify.

a) $4 : 11$ **b)** 6 to 13 **c)** 0.006

d) $0.\overline{255}$ **e)** 0.255 **f)** $0.2\overline{55}$

33. Estimate the value of the following irrational numbers without a calculator and round to the nearest hundredth.

a) $\sqrt{13}$ **b)** $\sqrt{21}$ **c)** $\sqrt{46}$

CHAPTER 2

Percent

2.1 The Percent of a Number

2.2 The Percent of Two Numbers

2.3 Discount, Tax, GST and PST

2.4 Commission Rates and Interest Rates

2.5 Per mil, Ppm and Ppb*

Chapter 2 TEST

* This section is optional. Teachers and students can decide whether they need to study it.

Chapter 2 Percent

2.1 The Percent of a Number

A percent can be described as a fraction with a denominator of 100. It is shown using a percent sign(%). For example, 79% means $\frac{79}{100}$. This number can also be called a percentage.

Changing Percentages to Decimals

To change a percentage to a decimal, move the decimal point of the number two digits to the left and then delete the % sign.

Example 1

Change the following percents to decimal form.

a) 58% **b)** 0.09% **c)** 104% **d)** 150.3%

▶ *Solution*:

a) 58%=0.58 **b)** 0.09%=0.0009

c) 104%=1.04 **d)** 150.3%=1.503

Changing Percentages to Fractions

To change a percentage to a fraction, delete the % sign and make a fraction with the percentage as the numerator and 100 as the denominator. Then simplify the fraction to its simplest form. For example, to convert 37.5% to a fraction, first delete the % sign. 37.5 then becomes the numerator of the fraction with a denominator of 100. The percentage is now $\frac{37.5}{100}$. Simplify this fraction:

$$\frac{37.5}{100}=\frac{37.5\times10}{100\times10}=\frac{375}{1000}=\frac{375\div125}{1000\div125}=\frac{3}{8}.$$

Example 2

Change the following percentages into fractions.

a) 83% **b)** 75% **c)** 25.5% **d)** 0.85%

▶ *Solution*:

a) $83\% = \frac{83}{100}$

b) $75\% = \frac{75}{100} = \frac{3}{4}$

c) $25.5\% = \frac{25.5}{100} = \frac{255}{1000} = \frac{51}{200}$

d) $0.85\% = \frac{0.85}{100} = \frac{85}{10\ 000} = \frac{17}{2000}$

Changing Decimals to Percentages

To change a decimal to a percentage, move the decimal point of the number two digits to the right and then add a % sign to the right of the new number.

Example 3

Change the following decimals to percentages.

a) 0.32 b) 0.07 c) 10.3 d) 1.0325

▶ *Solution*:

a) $0.32 = 32\%$ b) $0.07 = 7\%$ c) $10.3 = 1030\%$ d) $1.0325 = 103.25\%$

Example 4

Change the following repeating decimals to percentages and round to a tenth of a percent.

a) $0.\overline{39}$ b) $0.\overline{66}$ c) $0.4\overline{5}$ d) $1.30\overline{5}$

▶ *Solution*:

a) $0.\overline{39} = 39.3939\ldots\% \approx 39.4\%$

b) $0.\overline{66} = 66.66\ldots\% \approx 66.7\%$

c) $0.4\overline{5} = 45.55\ldots\% \approx 45.6\%$

d) $1.30\overline{5} = 130.55\ldots\% \approx 130.6\%$

Changing Ratios or Fractions to Percentages

Firstly, convert the fraction or ratio to decimal form, then move the decimal point two digits to the right and add a percent sign (%).

Example 5

Change the following fractions or ratios to percents.

a) $\frac{3}{4}$ b) $\frac{88}{100}$ c) 15 : 75 d) 97 to 99

▶ *Solution*:

a) $\frac{3}{4} = 0.75 = 75\%$, or $\frac{3}{4} = \frac{3 \times 25}{4 \times 25} = \frac{75}{100} = 75\%$

b) $\frac{88}{100} = 88\%$

c) $15 : 75 = \frac{15}{75} = \frac{1}{5} = 0.2 = 20\%$

d) $97 \text{ to } 99 = \frac{97}{99} = 0.9797\ldots \approx 97.98\%$

Practice

■Level A

1. Change the following percentages to decimals.

a) 88% **b)** 76% **c)** 35% **d)** 69%

e) 0.06% **f)** 0.58% **g)** 1.20% **h)** 4.06%

i) 207% **j)** 314% **k)** 515% **l)** 903%

2. Change the following percentages to fractions.

a) 78% **b)** 65% **c)** 48% **d)** 29%

e) 55% **f)** 75% **g)** 42% **h)** 30%

i) 35.5% **j)** 22.4% **k)** 15.5% **l)** 65.5%

3. Change the following decimals to percentages.

a) 0.64 **b)** 0.57 **c)** 0.96 **d)** 0.32

e) 0.05 f) 0.25 g) 0.15 h) 0.08

i) 11.23 j) 20.5 k) 19.05 l) 123.3

Level B

1. Change the following fractions or ratios to percentages.

a) $\frac{3}{8}$ b) $\frac{3}{4}$ c) $\frac{1}{16}$ d) $\frac{5}{16}$

e) $\frac{18}{37}$ f) $\frac{13}{15}$ g) $\frac{12}{27}$ h) $\frac{15}{35}$

i) 8 : 25 j) 2 : 108 k) 15.7 : 50 l) 45 : 95

2. Change the following decimals or percentages to fractions.

a) 2.037 b) 0.125 c) 15.85 d) 10.005

e) 145.05% f) 23.68% g) 106.07% h) 652.10%

i) 65% j) 0.025% k) 0.125% l) 1.0015%

3. Bob got a mark of $\frac{86}{100}$ on his first math test and a mark of $\frac{92}{100}$ on his second math test. What was the average percentage he earned in his math course so far?

4. Mindy bought some fabric that was 2.25 meters long. How could this be written as a fraction in metres?

5. Leo bought his coat on Boxing Day for $\frac{2}{3}$ off the original price. What percentage was taken off the price of the coat?

6. Tony pays income tax at the rate of 25% of his income. What fraction of Tony's income does he pay in income tax?

7. Caroline is working out a problem involving the fraction $\frac{3}{4}$. She needs to enter this into a calculator. How would she enter $\frac{3}{4}$ as a decimal on the calculator?

8. According to a survey, one hundred in eight hundred people said that they preferred a particular brand of cola. What is this ratio as a percentage?

9. Order the following numbers from the smallest to the largest: $\frac{7}{8}$, 0.87, 87% and 0.8.

10. Change "2.4 metres to 60 cm" to a whole number ratio.

11. Cut a 7-metre stem into 6 pieces.

a) Write the length of each piece as a fraction.

b) How long is each piece to the nearest hundredth?

12. If the number of girls is $\frac{4}{5}$ of the number of boys in a class, what percentage of the class are boys? Round your answer to the nearest hundredth of a percent.

13. Write the next three numbers based on the pattern.

90%, $\frac{4}{5}$, 0.7, ( ), ( ), ( )

Level C

1. Use "=", ">" or "<" to compare each pair of numbers.

   a) $\frac{2}{3}$ 0.66 b) a half 5% c) $\frac{9}{101}$ 0.09

2. The interest rate per year at a bank is 2.25%. John deposited \$5000 into his account for one year. How much interest did he obtain?

3. A rectangular field has a perimeter of 280 m. The ratio of the length to the width is 5 to 2. If the field can grow 16 kg of wheat per square metre, how many kilograms of wheat can the whole rectangular field grow?

2.2 The Percent of Two Numbers

Finding the percentage of one number out of another number

Example 1

What percent of 125 is 43?

▶ *Solution*: $\frac{43}{125}=\frac{43\times 8}{125\times 8}=\frac{344}{1000}=\frac{34.4}{100}=34.4\%$

Example 2

What percent is 55 out of 195? Round to the nearest hundredth of a percent.

▶ *Solution*: $\frac{55}{195}\approx 0.2821=28.21\%$

Example 3

What percent of 25 is 1500?

▶ *Solution*: $\frac{1500}{25}=60=\frac{6000}{100}=6000\%$

Finding a number from a percentage of another number

Example 4

What number is 47% of 87?

▶ *Solution 1*: $87\times 47\%=87\times 0.47=40.89$

▶ *Solution 2*: $87\times 47\%=\frac{87\times 47}{100}=40.89$

Example 5

What number is 0.5% of 420?

▶ *Solution*: $420\times 0.5\%=420\times 0.005=2.1$

Example 6

78 is 25% of which number?

▶ *Solution 1*: Assign x to the number. Use $x\times 0.25=78$ to find x.

$x=\frac{78}{0.25}=312$

The number is 312.

▶ *Solution 2*: $\frac{78}{25\%}=\frac{78}{\frac{1}{4}}=78\times4=312$

Example 7

3600 is 180% of which number?

▶ *Solution*: Assign x to the number, $3600=180\%\times x=1.8x$

$x=3600\div1.8=2000$

The number is 2000.

Practice

■Level A

1. a) What percent of 25 is 21?

b) What percent of 135 is 45?

c) What percent of 5 is 7?

d) What percent of 12 is 21?

e) What number is 25% of 30?

f) What number is 5% of 23?

g) What number is 85% of 120?

h) What number is 75% of 140?

2. a) 45 is 35% of which number?

b) 25 is 25% of which number?

c) 32 is 49% of which number?

d) 15 is 55% of which number?

e) What number is 80% of $150?

f) What number is 30% of $240?

g) What number is 15% of 225 kg?

h) What number is 50% of 1250 cm?

3. 30% of Number A is 100 more than Number B. If Number A is 1500, what is Number B?

4. If Number A is 20% less than Number B, how much more is Number B than Number A? Write the answer as a percentage.

■Level B

1. **a)** What percent of 125 is 88?

b) What percent of 188 is 45?

c) What percent of 225 is 56?

d) What percent of 264 is 32?

e) What number is 35% of 290?

f) What number is 0.45% of 190?

g) What number is 1350% of 380?

h) What number is 21.5% of 124?

2. 30% of 300 is 50% of which number?

3. 30% of Number A is 100 more than Number B. If Number B is 1500, what is Number A?

4. If 40% of a number is 8 more than 20% of the same number, what is 80% of this number?

5. A factory was set to produce 1000 machinery parts. They ended up producing 1200 machinery parts. By how much did the factory overproduce? Write the answer as a percentage.

6. 20% of a pile of sand is taken away, leaving 1120 tons. What was the original weight of the pile of sand?

7. Tony entered a math contest. He knew that he answered 52 questions correct and 13 questions wrong. What is Tony's mark as a percentage?

8. There is 20 g of water in Cup A and 25 g of water in Cup B. If 7 g of sugar is added into Cup A and 9 g of sugar is added into Cup B, which cup of water will be sweeter?

9. Order the numbers from the smallest to the largest.

a) $\frac{3}{4}$, 67%, 0.666

b) 3.75%, 37.5%, $\frac{4}{9}$

c) 128.59%, 1.27, 12.9%

d) 24.5, $\frac{1}{4}$, 25.1%

■Level C

1. The price of a certain television is \$2500. If it goes on a sale, the price will decrease by \$200. By what percent will the price go down when it goes on a sale?

2. Solve the following equations.

a) $(1-25\%)x=72$

b) $x-40\%x=5.04$

c) $x\div(1-40\%)=3.6$

d) $3x-25\%x=12$

3. After 10 tons of wheat are moved from Warehouse A to Warehouse B, the amount of wheat at Warehouse A is 40% of that of Warehouse B. We know that Warehouse A originally stored 110 tons of wheat. How many tons of wheat were originally stored at Warehouse B?

2.3 Discount, Tax, GST and PST

Discount

A **discount** is a reduction in the price of an item from its original price.

Discount = Original price × Discount rate

Sale price = Original price − Discount

Example 1

A hat's original price is \$99.00. What is the sale price of the hat during a 15% off-sale?

▶ *Solution*: $99.00-99.00\times 15\%=99.00\times 1-99.00\times 15\%=99.00\times(1-15\%)$
$=99.00\times 0.85=84.15$

The sale price of the hat is \$84.15.

Example 2

After a 25% discount, the sale price of a pair of shoes is \$110.00. What was the original price of the shoes?

▶ *Solution*: Let the original price be x. So we have the equation:

$x(1-25\%)=110.00$

$x=110.00\div(1-25\%)=110.00\div 0.75=146.67$

The original price of the shoes is \$146.67.

GST and PST

PST (Provincial Sales Tax) and **GST** (Goods and Services Tax) are the names used to describe the sales taxes in parts of Canada. In British Columbia, the rate of PST is 7% and the rate of GST is 5%.

Example 3

How much tax did Mary pay if she bought a pair of shoes that cost \$260.00 before tax (5% GST and 7% PST will apply)?

▶ *Solution*: $260.00\times(7\%+5\%)=260.00\times 12\%=260.00\times 0.12=31.20$

Mary paid \$31.20 in taxes.

Example 4

How much did Jack pay altogether when he bought a \$2580.00 suit with 12% tax?

▶ *Solution*:

$2580.00 \times (1+12\%) = 2580.00 \times (1+0.12) = 2580.00 \times 1.12 = 2889.60$

Jack paid \$2889.60.

Example 5

Tony wants to buy a pair of jeans. The original price is \$199.00 before a 20% off-sale. Including 5% GST and 7% PST, how much should he pay?

▶ *Solution*:

$199.00 \times (1-20\%) \times (1+5\%+7\%) = 199.00 \times 0.8 \times 1.12 = 178.30$

Tony should pay \$178.30.

Example 6

A dress was on sale at 25% off. Mindy paid \$235.20 including tax (both 5% GST and 7% PST were charged). What was the original price of the dress?

▶ *Solution 1*:

The price before discount: $235.20 \div (1-25\%) = 235.20 \div 0.75 = 313.60$

Before the tax: $313.60 \div (1+5\%+7\%) = 313.60 \div 1.12 = 280.00$

The original price of the dress was \$280.00.

▶ *Solution 2*:

Let the original price be x, so the equation to describe the problem will be:

$x(1-25\%)(1+5\%+7\%) = 235.20$

$$x = \frac{235.20}{(1-25\%)(1+5\%+7\%)} = \frac{235.20}{0.75 \times 1.12} = 280.00$$

The original price of the dress was \$280.00.

Practice

(For the following questions, assume GST is 5%, PST is 7%)

■Level A

1. How much GST and PST did Jenny pay when she bought a skirt that costs \$230.00?

2. How much GST and PST did Mary pay when she bought a hat that costs \$80.00?

3. Tony paid \$13.20 in total for a product, including GST and PST. How much did the product cost before taxes?

4. Ben bought a product and paid \$23.50 with GST and PST. How much did the product cost before taxes?

5. Fill in the following table and round your answers to the nearest cent.

Selling Price	Tax Rate	Quantity of Tax	Total Cost
\$400.00	12%(GST+PST)	\$48.00	\$448.00
\$550.00	GST(5%)		
\$800.00	PST(7%)		
\$590.00	15%		
	17%	\$10.00	
	12%		\$234.00
	15%	\$30.00	

■Level B

1. How much did Jack pay including both PST and GST when he bought a \$2580.00 suit?

2. How much did Tony pay including both PST and GST when he bought some furniture for \$5999.00?

3. Bobby bought a pair of shoes originally priced at \$139.99 in a 20% off-sale. How much did he pay without taxes?

4. Wilson bought a laptop originally priced at \$899.99 in a 10% off-sale. How much did he pay without taxes?

5. Mr. Smith bought a pair of jeans originally priced at \$39.99 for 25% off. How much did he pay with a tax rate of 12%?

6. Fill in the following table. Round the answers to the nearest cent.

Regular Price	Discount Rate	Amount of Discount	Tax Rate	Amount of Tax	Sale Price
\$259.99	10%	\$26.00	GST+PST(12%)	\$28.08	\$262.07
\$99.99	15%		10%		
\$300.00	0			\$36.00	
\$599.99		\$120.00		\$15.60	
\$199.99	12%		10%		
\$1350.00	0			\$162.00	

7. Leon bought a refrigerator at 20% off. He paid \$1299.00 including GST and PST. What was the original price of the refrigerator? How much GST and PST was charged?

8. A car priced at an original price of \$110 000 with a sales tax rate of 7% was sold at a sale price of \$111 815 including tax. How much of a discount did the car dealer give the customer?

9. Mike bought a suit at 25% off. He paid \$428.40 including GST and PST. What was the original price of the suit? How much GST and PST was charged?

■Level C

1. The price of a table is \$10 higher than that of a chair. If the price of the chair is 60% of the price of the table, what is the price of each?

2. A school had 700 students in 2011. 280 students graduated and left the school in July 2012, but the school received 350 new freshmen in September 2012. What is the percentage increase in the number of students at the school?

3. 80 ml of a 5% salt water solution is mixed with 20 ml of an 8% salt water solution. 10 mL of the mixed solution is poured away and 10 ml of pure water is added to the solution. What is the new concentration of the salt water solution?

4. Charlie bought a new set of teacups which cost him \$71.40, after PST (7%) & GST (5%) and 15% discount. Assume that the original price of the teacups was marked up from their wholesale price by 20%. Calculate the wholesale price of the set of teacups.

2.4 Commission Rates and Interest Rates

Commission

A **commission** is an amount of money earned from selling a product or providing a service, usually calculated as a percentage of the amount sold.

Commission=(Total amount of money)×(Commission rate)

Example 1

John, a car dealer, receives a 3% commission from sales he makes. Last year, he sold $40 000.00 worth of products. His commission was $2 500.00 per month. How much did he earn last year?

▶ *Solution*:

$40\ 000.00\times 3\%+2\ 500.00\times 12=40\ 000.00\times 0.03+2\ 500.00\times 12=31\ 200.00$

He earned $31 200.00 last year.

Simple Interest

The **principal** is the amount of money deposited or borrowed from a bank or another money lender.

Simple Interest is the amount of money charged for borrowing money or the amount of money paid for saving money. It is expressed as a percentage rate and the following is known as the Simple Interest formula.

$I = P\times R\times T$

(Simple Interest=principal×interest rate per year×time in years).

Example 2

Jean borrowed $2000.00 from a bank at an annual interest rate of 5%. How much interest should she pay ten years later?

▶ *Solution*:

$2000.00\times 5\%\times 10=1000.00$ She should pay $1000.00 in interest.

Compound Interest

Compound Interest is calculated by a different method from simple interest. Using compound interest, the interest gained each year is added to the principal, so the compound interest starts to earn its own interest the next year.

Example 3

\$2000 was invested at 4% compound interest rate for 3 years. Use a table to describe the investment in each year.

▶ *Solution*:

Year	\$ at start of year	Interest	\$ at end of year	Comments
1	2000	80	2080	$80=2000\times4\%$, $2000+80$ $=2000+2000\times4\%$ $=2000(1+4\%)=2080$
2	2080	83.20	2163.20	$83.20=2080\times4\%$, $2080+83.20=2080+2080\times4\%$ $=2080(1+4\%)=2000(1+4\%)^2$ $=2163.20$
3	2163.20	86.53	2249.73	$86.53=2163.20\times4\%$, $2163.20+86.53=2163.20+$ $2163.20\times4\%=2163.20(1+4\%)$ $=2000(1+4\%)^3=2249.73$

Example 4

Mary made an investment of \$15 000 at a bank with a compound interest rate of 5%. How much will she earn two years later?

▶ *Solution*:

$15\ 000\times(1+5\%)\times(1+5\%)-15\ 000=1537.5$

May will earn \$1537.5 in interest two years later.

Percent Increase and Percent Decrease

Percent Increase and Percent Decrease can be described as the change in percentage of an original amount.

$$\text{Percent Increase}=\frac{\text{Amount of Increase}}{\text{Original Amount}}$$

$$\text{Percent Decrease}=\frac{\text{Amount of Decrease}}{\text{Original Amount}}$$

Example 5

The price of a watch was increased from \$200 to \$250. By what percentage was the price increased?

▶ *Solution*:

$\frac{250-200}{200}=\frac{50}{200}=0.25=25\%$

The price was increased by 25%.

Example 6

The sale price of some computers last year was $1450 while the price of the same model of computers this year is $1100. By how much did the price decrease? Express your answer in dollars and as a percentage.

▶ *Solution*:

$1100-1450=-350$

$\frac{-350}{1450}\approx-0.2414=-24.14\%$

The price of the computers decreased by $350 or by 24.14%.

Example 7

Tom gets a 2.5% increase in his salary every year. His current salary is $3800.00 per month. How much monthly salary should he expect to get 4 years later?

▶ *Solution*:

$3800.00\times(1+2.5\%)^4=3800\times1.025^4=4194.49$

His salary should be $4194.49 per month 4 years later.

Practice

■Level A

1. Fill in the following table.

Total Sale	Rate of Commission	Amount of Commission
$265 000	2%	$5300
	3%	$195
$1 659 800	2%	
$239 850	0.85%	
$565 800		$16 974
	4%	$295
$294 800	2.5%	
$39 810	1.85%	

2. Fill in the following table using the simple interest formula.

Principal	Interest Rate Per Year	Years	Simple Interest	Total Amount
\$ 5000.00	5%	2	\$ 500.00	\$ 5500.00
\$ 45 500.00	3%	3		
	4%	4	\$ 424.00	
	15%	2		\$ 12 740.00
\$ 236 500.00	8%		\$ 94 600.00	
\$ 39 800.00		3	\$ 5970.00	
	12%	2		\$ 1520.00
\$ 34 500.00	6%		\$ 10 350.00	

■Level B

1. Fill in the following salary list for a car dealer company.

Name of Employee	Total Amount of Sale	Rate of Commission	Amount of Commission	Amount of Regular Salary	Amount of Salary
A	\$ 35 000.00	2%	\$ 700.00	\$ 2500.00	\$ 3200.00
B	\$ 85 000.00	3%		\$ 2800.00	
C	\$ 10 500.00	3%			\$ 3515.00
D		2%		\$ 3000.00	\$ 3117.40
E	\$ 76 000.00	3%		\$ 2500.00	
F	\$ 12 530.00	3%			\$ 3255.90
G		2%		\$ 2900.00	\$ 4204.00

2. Mr. Li invested \$ 15 000 at 5.5% compound interest rate for 3.5 years. Ms. Lu invested \$ 18 000 at 5% compound interest rate for 3 years. Who would earn more interest?

3. Bill received two job offers at two car dealerships. One has a base salary of \$ 2000 per month with a 5% commission. The other has a base salary of \$ 2500 with a 2% commission. Which one do you think is better?

4. Fill in the following table. Round the answers to the nearest hundredth.

Original Number	New Number After Changing	Amount of Change	Percent of Change
\$ 5000	\$ 5500	+ \$ 500	+10%
\$ 4800	\$ 5400		
\$ 5900	\$ 5100		
\$ 300		+ \$ 50	
\$ 2800			−5%
\$ 6200	\$ 4600		
\$ 600		+ \$ 40	

5. Fill in the following table based on Compound Interest formula.

Principal	Interest Rate Per Year	Years	Compound Interest	Total Amount
\$ 5000	5%	1		
\$ 4500	3%	2		
\$ 8500	4%			\$ 8840
	15%	2		\$ 1322.50
\$ 2500	8%		\$ 416	
\$ 3900		2	\$ 399.75	
	12%	2		\$ 6021.12

6. \$ 25 000 was invested at 8% compound interest rate for 5 years. Fill in the following table.

Year	\$ at start of year	Interest	\$ at end of year
1			
2			
3			
4			
5			

Level C

1. A school has 318 students, and there are 12% more girls than boys. How many female students are there at the school?

2. There are two boxes of apples, and the number of apples in Box B is 50% of that in Box A. If Sarah takes 25 apples from Box A and puts them in Box B, then the number of apples in Box B will be 80% of that in Box A. How many apples are in Box A now?

3. Mary has fruit candies and milk candies. The number of milk candies is 45% of the total number of candies. If she gets 16 more fruit candies, the milk candies will be 25% of the number of total candies. How many milk candies does Mary have?

4. Chris deposited \$250 000 on Jan 1, 2009 in a savings account that pays a compound interest rate of 1.15% at the end of every year. As a day job, Chris started to work at an auto dealership at the same time and earned a flat salary of \$2 000 per month, plus another 3.5% in commission for each car sold. Chris sells 7 cars every month at \$14 000 per car. At the end of 2012, how much will he have earned including salary, commission and interest?

2.5 Per mil, Ppm and Ppb*

Per mil

Another ratio similar to percentage is called **per mil**, where the denominator is 1000 instead of 100. The sign for per mil is written as "‰" and 1‰ is defined as one tenth of 1%.

$1‰=\frac{1}{1000}=0.001=0.1\%$

$1\%=10‰$

Common applications of per mil include the following:

- legal limits of blood-alcohol content for driving a road vehicle
- seawater salinity (e.g. the average salinity is 35‰)
- baseball batting averages

Example 1

In British Columbia, Canada, the legal limit of blood-alcohol content for driving is 0.05%. Convert it to per mil.

▶ *Solution*:

Since $1\%=10‰$; 0.05% can be converted to per mil by multiplying it by 10:

$0.05\%=(0.05\times10)‰=0.5‰$

Ppm & Ppb

While percentage (%) and per mil (‰) have numerous useful applications in real life, sometimes the concentration is so low that expressing it in the above 2 units becomes tedious due to the amount of zeros. Let's say we want to express the concentrations of a certain pollutant in air, and for every 1 m^3 of air, 1 cm^3 of this pollutant is present.

The concentration by volume is then calculated as follows:

$\frac{1\ cm^3}{1\ m^3}=\frac{1\ cm^3}{1\ 000\ 000\ cm^3}=0.000001=0.0001\%=0.001‰$

As we can see, a smaller unit for measuring concentration is desired.

Ppm, or parts per million, is a common measurement for trace concentration. As the term indicates, 1 ppm simply means 1 part for every 1 000 000 parts.

---

* This section is optional. Teachers and students can decide whether they need to study it.

Conversions between ppm and common notations are stated as follows:

1 ppm=0.000001=0.0001%=0.001‰

1%=10 000 ppm

1‰=1 000 ppm

Ppb, or parts per billion, is another common measurement for concentrations of a particular substance and 1 ppb is equivalent to one thousandth of 1 ppm. Due to their nature, the two units are mainly used in science and engineering. Some of the common applications include:

- Pollution levels in air, water, etc.
- Amounts of nutrients (e.g. This bottle of vitamin water contains X ppm of vitamin C)
- Concentrations of acids (e.g. chloric acid) in household chemicals

Example 2

Carbon Monoxide (CO) is a poisonous gas which is harmful to animals and humans, and it often exists in homes because of poor furnace maintenance. It is produced in small amounts when fossil fuels are burned. If not ventilated properly, the concentration can quickly rise to a dangerous level. In a guideline published by WorkSafeBC in May 2013, 25 ppm is the maximum limit for 8-hour exposure. Convert 25 ppm to a percentage.

▶ *Solution*:

Since 1 ppm=0.0001%, 25 ppm=0.0001%×25=0.0025%

All homes should be equipped with a Carbon Monoxide and smoke detector.

Practice

■Level A

Convert the following concentrations to ppm and ppb.

a) 1.1%　　**b**) 0.5%　　**c**) 100%　　**d**) 19.9%

e) 7%　　**f**) 35‰　　**g**) 0.3‰　　**h**) 20‰

i)10. 3‰ **j**)8. 81‰

■Level B

Convert the following concentrations to percentage and per mil.

a)25 ppm **b**)2 ppm **c**)0. 5 ppm **d**)750 ppm

e)15 000 ppm **f**)580 ppb **g**)29 ppb **h**)3 ppb

i)0. 4 ppb **j**)6000 ppb

■Level C

Cadmium is a heavy metal that can cause cancer. In a table published by WorkSafeBC, the exposure limit for cadmium is 0. 01 mg/m^3. Convert it to ppb, write the final answer in scientific notation and round it to the nearest tenth. Assume the density of cadmium is 8. 65 mg/cm^3.

Chapter 2 TEST

1. Change the following fractions or ratios to percents.

a) $\frac{3}{7}$ **b)** $\frac{3}{9}$ **c)** $\frac{3}{16}$ **d)** $\frac{5}{8}$

e) $\frac{15}{32}$ **f)** $\frac{27}{12}$ **g)** 14 : 60 **h)** 42 : 168

2. Change the following decimals or percents to fractions.

a) 2.659 **b)** 0.375 **c)** 15.2 **d)** 20.05

e) 142.56% **f)** 21.6% **g)** 216.77% **h)** 659.23%

3. Cindy got $\frac{92}{100}$ on the first science test and $\frac{45}{60}$ on her second science test. What was the average percentage she earned in her science course so far if both tests weighted same?

4. Kevin bought a book at $\frac{1}{8}$ off the original price. What percentage was taken off the price of the book?

5. Solve.

a) What percent of 250 is 32? **b)** What percent of 132 is 58?

c) What number is 48% of 3000?

d) What number is 0.15% of 12?

e) What number is 2350% of 280?

f) What number is 41.5% of 1300?

6. 20% of 800 is 45% of which number?

7. If 35% of a number is 12 more than 18% of the same number, what is 65% of this number?

8. Solve for x.

a) $(1-35\%)x=26$

b) $x-24\%x=4.2$

c) $x\div(1-65\%)=3.2$

d) $3x-85\%x=10$

9. Fill in the following table and round the answers to the nearest cent.

Regular Price	Discount Rate	Amount of Discount	Tax Rate	Amount of Tax	Sale Price
\$49.99	10%	\$5.00	GST+PST(12%)		
\$199.99	15%		7%		
\$200.00	0%			\$14.00	
\$499.99		\$199.99		\$25.00	

10. Jacky paid \$671.99 to buy a camcorder at a store that was 25% off. What was the original price of the camcorder? How much GST (5%) and PST (7%) were charged?

11. The price of a math book is $20 higher than that of a science book. If the price of the science book is 85% of the price of the math book, what is the price of each book?

12. Fill in the following table using the simple interest formula.

Principal	Interest Rate Per Year	Years	Simple Interest	Total Amount
$45 500.00	3%	3		
	4%	4	$424.00	
$236 500.00	8%		$94 600.00	
$34 500.00	6%		$10 350.00	

13. Fill in the following salary list for a company.

Name of Employee	Total Amount of Sale	Rate of Commission	Amount of Commission	Amount of Regular Salary	Amount of Salary
A	$45 000.00	2.5%		$2500.00	$3625.00
B	$85 000.00	3%		$2800.00	
C	$76 000.00	3%		$2500.00	
D		2%		$2900.00	$4204.00

14. Fill in the following table based on the compound interest formula.

Principal	Interest Rate Per Year	Years	Compound Interest	Total Amount
$2500	5%	1		
$8200	4.5%			$8954.61
$3200	8%		$256	
	12%	2		$6021.12

CHAPTER 3

Ratios, Proportions, Scales and Rates

3.1 Ratios, Proportions and Scales

3.2 Rates and Unit Rates

3.3 More Application Problems Regarding Rates and Ratios

Chapter 3 TEST

Chapter 3 Ratios, Proportions, Scales and Rates

3.1 Ratios, Proportions and Scales

Ratios

The quotient of two numbers is called a **ratio.** A ratio can be written as $a : b$ or a to b. A ratio can also be written as a fraction.

Ratios can be simplified by dividing the denominator and the numerator by their GCF(Greatest Common Factor).

Example 1

Simplify the ratio 7 : 21.

▶*Solution*:

$\frac{7}{21}=\frac{7\div 7}{21\div 7}=\frac{1}{3}$

The simplified ratio is 1 : 3.

Equivalent ratios can be formed by multiplying or dividing the terms by the same non-zero number. For example:

$3:5=3\times(-2):5\times(-2)=-6:-10$

Proportions

A **proportion** is an equation that states that one ratio is equal to another ratio. Proportions can be expressed with two or more ratios by using equal signs.

Solving Proportion Equations

If a proportion includes an unknown term (x), it can be solved by equating the cross-product.

Example 2

Find the unknown term in the following proportion: $\frac{x}{330}=\frac{7}{21}$.

▶ *Solution*: The cross-product of the terms are equal.

$\frac{x}{330}=\frac{7}{21}$, $\frac{x}{330} \times \frac{7}{21}$

$21\times x=(330)\times(7)$, $x=\frac{330\times 7}{21}=110$

Example 3

Find the unknown term:

$12:x=2:12$

▶ *Solution*: The product of the two inside terms is equal to the product of the two outside terms.

$12:x=2:12$, $2x=(12)(12)$, $x=\frac{144}{2}=72$

Map Scales

Map scales are an application of ratios that compare distances on a map to actual distances.

Example 4

The measurement of a distance on a map from City A to City B is 4 cm and the map's scale is 1 : 1 000 000. Find the actual distance between these two cities.

▶ *Solution*: Let x be the actual distance between these two cities (in km).

We set up a proportion:

$1:1\ 000\ 000=4:x$ or $\frac{1}{1\ 000\ 000}=\frac{4}{x}$

$1\times x=4\times 1\ 000\ 000$, $x=4\ 000\ 000$ cm$=40\ 000$ m$=40$ km

The distance between the two cities is 40 km.

Example 5

If the ratio of $a:b$ is $5:7$, and $a=45$, find the value of b.

▶ *Solution*: Set up a proportion and substitute $a=45$, then $45:b=5:7$.

Since the product of the two outside terms is equal to that of the two inside terms, we can set up the equation as follows:

$5b=7\times 45=315$

$b=63$

Example 6

If the ratio of $a:b$ is $3:4$ and the ratio of $b:c$ is $6:11$, what is the ratio of $a:c$?

▶ *Solution*:

Since equivalent ratios can be formed by multiplying each value in a ratio by the same number, $a:b$ and $b:c$ can be rewritten as follows:

$a:b=3:4=(3\times3):(4\times3)=9:12$

$b:c=6:11=(6\times2):(11\times2)=12:22$

Therefore, $a:b:c=9:12:22$, and $a:c=9:22$.

Example 7

The ratio of $a:b$ is $6:13$ and the ratio of $b:c$ is $39:11$. If $a=36$, find the value of c.

▶ *Solution*: We first need to find the ratio $a:b:c$.

$a:b=6:13=(6\times3):(13\times3)=18:39$

$b:c=39:11$

Therefore, $a:b:c=18:39:11$

$a:c=18:11=36:c$

$8c=11\times36=396$

$c=22$

Practice

■Level A

1. Write down three equivalent ratios for each of the following ratios.

a) $3:7$ **b**) $4:9$ **c**) $5:13$ **d**) $6:25$

e) 5 to 19 **f**) 4 to 25 **g**) 5 to 15 **h**) 7 to 23

i) $\frac{7}{11}$ **j**) $\frac{15}{35}$ **k**) $\frac{16}{22}$ **l**) $\frac{13}{28}$

m) $\frac{17}{49}$ **n**) $\frac{13}{27}$ **o**) $\frac{15}{55}$ **p**) $\frac{25}{125}$

2. Find the unknown terms and round the answers to the nearest tenth if necessary.

a) $3:7=x:56$ **b)** $5:x=6:54$ **c)** $3:5=77:x$

d) 5 to $19=30$ to x **e)** x to $24=10$ to 6 **f)** 5 to $15=x$ to 320

g) $\frac{x}{176}=\frac{7}{11}$ **h)** $\frac{5}{15}=\frac{x}{121}$ **i)** $\frac{x}{63}=\frac{27}{99}$

j) $\frac{21}{x}=\frac{33}{210}$ **k)** $\frac{42}{x}=\frac{32}{21}$ **l)** $\frac{12}{85}=\frac{20}{x}$

■Level B

1. Write down three equivalent ratios based on the following ratios.

a) $4:3:7$ **b)** $5:2:6$ **c)** $8:7:9$

d) 7 to 5 to 19 **e)** 6 to 7 to 13 **f)** $1:13:15$

g) $\frac{0.7}{1.1}$ **h)** $\frac{0.5}{1.2}$ **i)** $\frac{0.8}{0.25}$

j) $\frac{2.1}{89}$ **k)** $\frac{2.5}{1.25}$ **l)** $\frac{1.8}{10.05}$

2. Find the unknown terms and round the answers to the nearest tenth if necessary.

a) $\frac{x}{6}=\frac{5}{15}=\frac{y}{30}$ **b)** $\frac{x}{9}=\frac{y}{25}=\frac{45}{120}$ **c)** $\frac{99}{9}=\frac{y}{25}=\frac{45}{x}$

d) $\frac{2}{6}=\frac{x}{25}=\frac{y}{13}$ **e)** $\frac{y}{6}=\frac{x}{25}=\frac{39}{13}$ **f)** $\frac{7}{y}=\frac{11}{x}=\frac{33}{210}$

g) $\frac{2}{8}=\frac{y}{25}=\frac{30}{x}$ **h)** $\frac{32}{y}=\frac{25}{x}=\frac{8}{40}$ **i)** $\frac{77}{y}=\frac{21}{x}=\frac{23}{210}$

3. Find the unknown terms and round the answers to the nearest tenth if necessary.

a) $3:7=x:56=y:49$ **b)** $14:6=52:x=y:9$ **c)** $32:y=43:35=x:25$

d) $x:4=16:y=24:36$ **e)** $23:x=56:y=36:108$

f) 5 to 9 = 30 to x = y to 27 **g)** x to 5 = 15 to y = 25 to 75

4. There are 17 boys out of 35 students in a table tennis tournament.

a) What is the ratio of boy players to total players?

b) What is the ratio of girl players to total players?

c) What is the ratio of girl players to boy players?

5. The distance on a map between two cities with a map scale of 1 : 10 000 000 is 3.5 cm. What is the actual distance between the two cities? (in km)

6. Two rectangular prisms A and B have the same base area. The ratio of the height of rectangular Prism A to rectangular Prism B is 7 : 11. If the volume of rectangular Prism B is 143 m^3, what is the volume of rectangular Prism A?

7. If I use square bricks with a side length of 40 cm to cover a walkway, it would take 9000 bricks to completely cover the walkway. If I change the bricks to squares with a side length of 30 cm, how many bricks would I need?

8. If cutting a steel bar into 4 sections takes 24 minutes, how many minutes do you need to cut the same steel bar into 7 sections?

9. Find the ratio of $a : c$ when given the following ratios.

a) $a : b = 5 : 1$; $b : c = 3 : 4$

b) $a : b = 5 : 12$; $b : c = 18 : 7$

10. Find the value of c when given the following ratios and values.

a) $a : b = 4 : 9$; $b : c = 6 : 7$; $a = 24$

b) $a : b = 8 : 15$; $b : c = 12 : 5$; $a = 64$

■Level C

1. There are two containers with the same carrying capacity. They are both filled with salt water. The ratio of water to salt in the first container is 2 : 3. The ratio of water to salt in the second container is 3 : 4. If I mix the salt water in the two containers, what is the new ratio of water to salt in the newly mixed salt water?

2. The ratio of the weights of two bags of sugar is 4 : 1. If I take 10 g of sugar from Bag A and put it into Bag B, the ratio of the two bags of sugar will become 7 : 5. What is the combined weight of the two bags of sugar?

3. Larry, Moe and Harry are 3 independent contractors. For a specific renovation task, it takes 40, 35 and 50 days for Larry, Moe and Harry to complete, respectively. Larry works five days a week, Moe only works on Mondays, Wednesdays, Fridays and Sundays, and Harry only works on Tuesdays, Thursdays and Saturdays. How many weeks does it take to complete the task if all 3 of them are put to work at once? Round the answer to the nearest whole number.

4. Donna is making a cheesecake for a party. The recipe for the cheesecake includes the following:

- 2 tablespoons of melted butter
- 4 eggs
- $1\frac{1}{2}$ cups of white sugar
- $\frac{3}{4}$ cup of milk

a) While shopping for the ingredients at a local grocery store, she found that only 3 eggs were available for purchase. To keep the same proportions, how much of the other ingredients are needed? Give the answer in exact values.

b) Suppose sugar costs \$2.82/bag and each bag weighs 5 lbs. Calculate the cost of sugar. Ignore taxes and round the answer to the nearest cent. (1 lb=454 g; 1 cup=200 g)

3.2 Rates and Unit Rates

Rates

A **Rate** is the comparison of two quantities with different units. Rates are quotients that can be described as one quantity divided by another.

Example 1

Bob rode a bike 30 000 metres in 120 minutes. What is his per-minute rate of speed when riding the bike?

▶*Solution*: $30\ 000 \div 120 = 250$ metres/minute

His riding rate is 250 metres/minute.

Unit Rates

A **Unit Rate** is the rate per unit. A unit rate is often used to describe price or the speed of an object.

Example 2

Linda paid \$7.50 to buy 3 small watermelons. How much did each watermelon cost?

▶*Solution*: $7.5 \div 3 = 2.5$

The price of one watermelon was \$2.50.

Example 3

In 5 hours, 2.5 litres of gas were used. How long will it take to use 15 litres (L)?

▶*Solution* 1 (proportion):

Assume x is the time it takes to use 15 L.

We set up the proportion:

$\frac{5}{2.5} = \frac{x}{15}$, $2.5x = 5 \times 15$, $x = \frac{5 \times 15}{2.5} = 30$

▶*Solution* 2 (unit rate—per-hour):

Since 2.5 L of gas were used in 5 hours, the per-hour rate is:

$\frac{2.5\text{ L}}{5\text{ h}} = 0.5\text{ L/h}$

$15\text{ L} \div 0.5\text{ L/h} = 30\text{ h}$

▶*Solution* 3 (unit rate—per-litre):

The per-litre rate is:

$\frac{5\ \text{h}}{2.5\ \text{L}}=2\ \text{h/L}$

$15\ \text{L}\times 2\ \text{h/L}=30\ \text{h}$

It takes 30 hours to use 15 litres of gas.

Example 4

Every morning Steven jogs a distance of 6.75 km in 45 minutes. What is his rate in meters per minute?

▶ *Solution*:

To get a rate in meters per minute, we must first convert kilometres to metres, and then divide the distance by the time elapsed.

$\frac{6.75\ \text{km}}{45\ \text{min}}=\frac{6\,750\ \text{m}}{45\ \text{min}}=150\ \text{m/min}$

Example 5

Tony prepared for a math contest by doing math drills. During the first 15 minutes, he finished 3 questions. In the next half an hour, he completed another 7. In the last hour, he did another 12. What was his rate of doing drills per hour?

▶ *Solution*:

Since we are asked for the rate by the hour, this is the unit we are going to use.

$\frac{3\ \text{Qs}+7\ \text{Qs}+12\ \text{Qs}}{15\ \text{min}+30\ \text{min}+60\ \text{min}}=\frac{22\ \text{Qs}}{0.25\ \text{hr}+0.5\ \text{hr}+1\ \text{hr}}=\frac{22\ \text{Qs}}{1.75\ \text{hr}}\approx 13\ \text{Q/hr}$

Tony did 13 questions per hour.

Practice

■Level A

1. Find the unit rate in the following questions.

a) 6000 pages of copy paper in 10 boxes.

b) 50 phone calls in 7 days.

c) 480 km traveled in 8 hours.

d) 250 calories in 30 minutes.

e) Typed 420 words in 7 min.

f) $15 600 in 6 months.

g) 10 people for 40 slices of pizza.

h) 6 case workers for every 100 clients.

i) $2550 to purchase 3 suits.

j) 4 tanks of gas to cut 16 acres of lawn.

2. Answer the following questions.

a) If 150 kg of rock is mixed with 250 kg of sand, how much rock is needed for 40 kg of sand?

b) If 15 tickets cost $375, how much do 7 tickets cost?

c) If 1.5 cm represents 500 km on a map, what distance does 2.9 cm represent?

d) How far can you travel in 1.5 hours at a running speed of 122 meters/min?

e) At 150 words in 4 minutes, how many words can be printed in 1 hour?

f) At what rate did Bob's monthly salary increase if his monthly salary rose from $2900 to $3100?

■Level B

1. John can build a 3-metre-long fence in 5 hours.

a) What is his working speed?

b) If his garden's length is 5 metres and the width is 4 metres, how long will he need to finish a fence to enclose his entire garden?

c) If the length and width of his garden are both increased by 1.5 metres, how long will he need to finish building the fence? Round your answer to the nearest minute.

d) If his working speed increases by 20%, and he works 8 hours each day, how many days will it take to finish his entire garden?

2. Five people are competing in a typing contest. The following table lists their results. Fill in the missing information.

Speed=(Quantity of Typed Words−Quantity of Wrong Words)÷Typing Minutes

Name	Quantity of Typed Words	Time Taken in Minutes	Quantity of Wrong Words	Quantity of Correct Words	Speed (words/min)	Order in the Contest
Mary	1000	25	25			
Tony	1000	25	250			
Arleen	1000		100		30	
Martin	1000	25	0			1
Wilson	1000		210		33	

3. Determine two numbers that are in a ratio of 9 : 7 if the sum of the two numbers is 48.

4. On June 7th, US $ 35 000 (US dollars) could be exchanged to C $ 35 357. 65 (Canadian dollars). What was the exchange rate?

5. Sound travels in water at a speed of 1460 m/s.

a) How far does sound travel in 2. 8 seconds in water?

b) How long does sound take to cross a distance of 10 000 metres?

6. Tom plans to go to US for shopping, and he exchanged C $ 2000 (Canadian dollars) to US $ 1979. 77 (US dollars). What was the exchange rate?

7. Decide which is a better deal by completing the table.

Items	Deal A	Deal B	Decide A or B
Apples	6 kg for $ 22. 08	10 kg for $ 29. 90	
Rice	5 kg for $ 42. 00	25 kg for $ 68. 75	
Printing job	6 hours for $ 86	20 hours for $ 260	
Discharging Cargo job	8 hours for $ 144	40 hours for $ 700	
Socks	2 dozens for $ 33. 60	25 pairs for $ 40	

8. Building workers make 96 tons of concrete building material with cement, sand, and gravel. The ratio of cement to sand to gravel is 2 : 3 : 5. How many tons of each ingredient do the workers need to make the building material?

9. A triangular copper wire has a perimeter of 84 cm. The ratio of the three sides of the triangle is 3 : 4 : 5. What is the length of each side of the triangle?

10. The average of three numbers is 88. Their ratio is 4 : 7 : 13. What are the three numbers?

11. A medicinal liquid is made by mixing medicine and water at a ratio of 3 : 400.

a) If I need 1612 g of medicinal liquid, how many grams of medicine will I need?

b) If I use 60 kg of water, how many grams of medicine will I need?

c) If I use 48 g of medicine, what is the maximum amount of medicinal liquid that can be made? Show your answer in exact kg.

12. Aron rides a bicycle to school. The distance between his home and the school is 4400 m. It takes him 19 minutes to get to school. What is his speed in kilometres per hour?

13. Jeff's typing speed is 57 wpm(words per minute). If his term paper contains 1250 words, how long will it take him to type it?

Level C

1. The ratio of the number of people in a singing club to the number of people in a dancing club is 3 : 2. If 10 people move from the singing club to the dancing club, the ratio of the number of people would change to 7 : 8. How many people are currently in the singing club?

2. The ratio of the merchandise stored in two warehouses A and B is 4 : 3. If the workers move 8 tons of merchandise from Warehouse A to Warehouse B, the ratio of the merchandise in the two warehouses will change to 4 : 5. How many tons of merchandise are stored in each of the warehouses?

3. The ratio of Kevin's walking speed to Robin's walking speed is 7 : 5. Kevin and Robin walk towards each other from Place C and Place D respectively at the same time. They meet each other after 0.5 hours. If Robin had walked in the opposite direction instead of towards Kevin, how much time would it have taken for Kevin to catch up with Robin?

4. A glass jar contains 50 chocolate chip cookies. Daniel eats 5 cookies every 3 days, Dennis eats 10 cookies every week, and Donna eats half a cookie every day. How many days will it take for them to finish 2 jars of cookies? Round the answer to the nearest whole number.

3.3 More Application Problems Regarding Rates and Ratios

A few more examples are given in this section to describe different application problems in our daily life.

Example 1

According to the 2011 Census of Canada, in Toronto the population ratio of South Asians to Latin Americans is 4 to 1, and the ratio of Latin Americans to First Nations is 6 to 1. If all three ethnic groups total 409 572, what is the population of Latin Americans?

▶ *Solution*: First of all, let's find the ratio between these three ethnic groups.

South Asians : Latin Americans $=4:1=(4\times6):(1\times6)=24:6$

Latin Americans : First Nations $=6:1$

Therefore, South Asians : Latin Americans : First Nations $=24:6:1$

Population of Latin Americans $=409\ 572\times\frac{6}{24+6+1}=409\ 472\times\frac{6}{31}=79\ 272$

Example 2

Jarvis is downloading an e-mail attachment. It takes him 32 seconds to download the 4 MB file. What is the rate of downloading in KB per second? Assume 1 MB=1000 KB.

▶ *Solution*: Before calculating the rate, we need to convert MB to KB first.

$\frac{4\text{ MB}}{32\text{ s}}=\frac{4000\text{ KB}}{32\text{ s}}=125\text{ KB/s}$

Example 3

Bruce is driving on a highway at a speed of 75 km/hr. How far will he travel if he drives for two and a half hours?

▶ *Solution*: Since he travels 75 kilometres every hour, the distance he will travel in two and a half hours is as follows:

$75\ \frac{\text{km}}{\text{hr}}\times2.5\text{ hr}=187.5\text{ km}$

Example 4

The average speed of a baseball pitch for a high school student is about 80 mph (roughly 128.75 km/hr). If the distance between the pitcher's mound and home plate is 18.39 m, how long does it take a baseball to reach home plate after being thrown?

▶ *Solution*: We must first convert the speed from kilometres per hour to metres per second.

$$128.75\ \frac{\text{km}}{\text{hr}}\times\frac{1000\ \text{m}}{1\ \text{km}}\times\frac{1\ \text{hr}}{60\ \text{min}}\times\frac{1\ \text{min}}{60}\text{s}=35.76\ \text{m/s}$$

If it takes the baseball one second to travel 35.76 m, then the time it takes to travel 18.39 m is:

$$\frac{18.39\ \text{m}}{35.76\ \text{m/s}}=0.51\ \text{s}$$

Practice

■Level A

1. The distance on a map between Ottawa and Quebec is 14 cm. On the map, every 3.8 cm represents 100 km. Find the actual distance between the two cities.

2. Steve is drawing a map as an assignment for his Social Studies course. If he uses a scale of 1∶10 000 000, and the actual distance between Edmonton and Calgary is 272.7 km, how far away should the two cities be placed on the map?

3. Jennifer works at a travel agency that sells flight tickets, package tours and cruises. In a particular month, the ratio of flight tickets sold to tours sold is 12∶25, and the ratio of tours sold to cruises sold is 2∶1. If she made 297 sales in total, how many cruises did she sell during that month?

4. According to a 2011 Census of Canada, in Vancouver the ratio of native English speakers to native French speakers is 100 : 3, and the ratio of native French speakers to those whose first language is neither English nor French is 1 : 30. If all three groups total 579 000 people, how many people are there whose first language is neither English nor French?

5. Jim goes on a road trip with his friends during the summer. For the first 250 km they drive at 80 km/hr. For the second part it takes them 3.5 hours to travel 300 km. For the last 150 km they drive at 95 km/hr. What is their average driving speed?

Level B

1. Cindy works at a snack shop that sells cream puffs. On a particular day her sales revenue is \$129 in the first two hours. For the next three hours it is \$206. For the last hour it is \$55. What is her average revenue per hour?

2. A calorie is a measurement of energy, and 1 calorie is the energy required to heat up 1 g of water by 1℃. How much energy is required to heat up 250 mL of room temperature water (at 25℃) to 99℃? Assume 1 mL of water weighs 1 g.

3. A knot (kt) is a common measurement for speed when sailing. One knot is equal to one nautical mile per hour. If a sailboat travels at 5.5 knots for two and a half hours, what is the distance traveled in kilometres? Assume 1 nautical mile=1852 m.

4. A car has a fuel consumption of 32 mpg (miles per gallon). If the tank's size is 11.1 gallons, how far can the car travel with a full tank of fuel?

5. A printer has a printing speed of 27 ppm (pages per minute). How many pages can it print in 5 minutes and 40 seconds?

■Level C

1. Joseph is planning to buy a condominium. If the down payment is \$14 000 and he saves 30% of his monthly salary (\$2800) each month, how long will it take him to save up enough to make the down payment?

2. The speed of light is approximately 300 000 km/s. If the distance between the Sun and the Earth is 149 597 871 km, how long does it take sunlight to reach the Earth from the Sun?

Chapter 3 TEST

1. Find the missing terms and round the answers to the nearest tenth if necessary.

a) $2:9=x:63$ **b)** $7:4=77:x$ **c)** 6 to $13=36$ to x

d) 5 to $19=15$ to x **e)** $\frac{x}{142}=\frac{7}{12}$ **f)** $\frac{15}{16}=\frac{x}{300}$

2. Find the missing terms and round the answers to the nearest tenth if necessary.

a) $\frac{x}{9}=\frac{36}{81}=\frac{y}{6}$ **b)** $\frac{x}{7}=\frac{y}{28}=\frac{26}{91}$ **c)** $\frac{42}{12}=\frac{y}{36}=\frac{45}{x}$

d) $x:8=4:y=24:36$ **e)** $23:x=46:y=36:108$

f) 6 to $9=30$ to $x=y$ to 45 **g)** x to $15=10$ to $y=25$ to 75

3. There are 25 boys out of the 45 players on a badminton team.

a) What is the ratio of boy players to total players?

b) What is the ratio of girl players to total players?

c) What is the ratio of girl players to boy players?

4. The distance on a map between two cities with a map scale of 1 : 1 000 000 is 4.9 cm. What is the actual distance between the two cities?

5. Find the unit rate in the following questions.

a) Typed 350 words in 5 minutes.

b) 8 people for 40 slices of pizza.

c) $16 800 in 8 months.

d) 6 case workers for every 120 clients.

e) $2670 to purchase 3 suits.

f) 8 tanks of gas to cut 32 acres of lawn.

6. Answer the following questions.

a) If 20 kg of rock is mixed with 16 kg of sand, how much sand is needed to mix with 420 kg of rock?

b) If 30 tickets cost $390, how much do 17 tickets cost?

c) If 2.5 cm represents 400 km on a map, what distance does 3.7 cm represent?

d) How far can you travel in 2.5 hours at a walking speed of 75 metres/min?

7. Robin can build a 4-metre-long fence in 5.5 hours.

a) What is his working speed(per hour)?

b) If the garden's length is 4.5 metres and the width is 2.5 metres, how much time will he need to finish the fence to enclose his entire garden?

c) If the length and width of his garden are both increased by 1.5 metres, how much time will he need to finish building the fence? Round your answer to the nearest minute.

d) If his working speed goes up by 25%, and he works 8 hours each day, how many days will it take to finish his job?

8. Fill in the spaces in the table for the word typing contest.

Speed =(Quantity of Typed Words − Quantity of Wrong Words) ÷ Typing Minutes

Name	Quantity of Typed Words	Time Taken in Minutes	Quantity of Wrong Words	Quantity of Correct Words	Speed (words/min)	Order in the Contest
Jerry	1500	20	20			
Cindy	1500	20	125			
Linda	1500		125		30	
Alyssa	1500	20	0			
Kevin	1500		180		33	

9. Determine 2 numbers if their ratio is 19 : 17 and the sum of the two numbers is 72.

10. Kathleen plans to go to US for shopping. She exchanged C$3500 (Canadian dollars) to US$3328.66 (US dollars). What is the exchange rate?

11. In Canada, 38% of the population resides in Ontario. The ratio of the population in Ontario to the population in Quebec is 19 : 12, and the ratio of the population in Quebec to the population in Saskatchewan is 8 : 1. Assume the total population of Canada is 33 477 000. What is the population of Saskatchewan?

12. On a map, the straight line distance between Vancouver and Portland is 5 cm. The map is drawn to a scale of 1 : 10 000 000. Assume this is the driving distance. (Round the answer to the nearest tenth.)

a) How long does it take to drive (in hours) from Vancouver to Portland if the first 3/4 of trip is done at 70 km/hr, and the last 1/4 is done at 110 km/hr?

b) Suppose the car has a fuel consumption of 45 mpg (miles per gallon). How many litres of gasoline are required for a round trip? (1 mile=1.609 km; 1 gallon=3.785 L)

CHAPTER 4

Multiplication and Division of Fractions and Integers

4. 1 Multiplication and Division of Fractions

4. 2 Multiplication and Division of Integers

4. 3 Order of Operations

Chapter 4 TEST

Chapter 4 Multiplication and Division of Fractions and Integers

4.1 Multiplication and Division of Fractions

Multiplication of Fractions

The basic steps of **Multiplying Fractions** are:

1. Simplify the fractions by reducing a numerator with a denominator.
2. Multiply the numerators and multiply the denominators.
3. Simplify the final answer if possible.

Example 1

$\frac{12}{16}\times\frac{25}{35}$

▶ *Solution*:

$\frac{12}{16}\times\frac{25}{35}=\frac{3}{4}\times\frac{5}{7}=\frac{15}{28}$

$\frac{12}{16}$ can be simplified to $\frac{3}{4}$ by dividing both the numerator and denominator by 4, and $\frac{25}{35}$ can be simplified to $\frac{5}{7}$.

Example 2

$\frac{9}{13}\times\frac{26}{27}$

▶ *Solution*:

$\frac{9}{13}\times\frac{26}{27}=\frac{1}{1}\times\frac{2}{3}=\frac{2}{3}$

9 and 27 in these fractions can be simplified by dividing each by 9. 13 and 26 can be simplified by dividing each by 13.

Example 3

$\frac{17}{19}\times\frac{26}{78}\times\frac{38}{51}$

▶ *Solution*: $\frac{17}{19}\times\frac{26}{78}\times\frac{38}{51}=\frac{17}{19}\times\frac{\cancel{26}^{1}}{\cancel{78}_{3}}\times\frac{38}{51}=\frac{17}{\cancel{19}_{1}}\times\frac{1}{3}\times\frac{\cancel{38}^{2}}{51}=\frac{\cancel{17}^{1}}{1}\times\frac{1}{3}\times\frac{2}{\cancel{51}_{3}}$

$=\frac{1}{1}\times\frac{1}{3}\times\frac{2}{3}=\frac{2}{9}$

Example 4

A rabbit and a tortoise compete in a 100-metre race. The rabbit runs $13\frac{1}{2}$ times faster than the tortoise. If the speed of the tortoise is $\frac{2}{5}$ m/min, what is the speed of the rabbit?

▶ *Solution*: $\frac{2}{5}\times13\frac{1}{2}=\frac{2}{5}\times\frac{27}{2}=\frac{54}{10}=\frac{27}{5}=5\frac{2}{5}$.

The speed of the rabbit is $5\frac{2}{5}$ m/min.

Division of Fractions

An easy method to divide fractions is to flip the second fraction (this is called the reciprocal), then multiply the reciprocal.

Example 5

$\frac{11}{15}\div\frac{22}{35}$

▶ *Solution*: $\frac{11}{15}\div\frac{22}{35}=\frac{11}{15}\times\frac{35}{22}=\frac{1}{3}\times\frac{7}{2}=\frac{7}{6}=1\frac{1}{6}$

Multiplication and Division of Mixed Fractions

Change mixed fractions to improper fractions, then use the standard method to solve it.

Example 6

$1\frac{1}{2}\times2\frac{1}{3}\times1\frac{1}{4}$

▶ *Solution*: $1\frac{1}{2}\times2\frac{1}{3}\times1\frac{1}{4}=\frac{3}{2}\times\frac{7}{3}\times\frac{5}{4}=\frac{3\times7\times5}{2\times3\times4}=\frac{7\times5}{2\times4}=\frac{35}{8}=4\frac{3}{8}$

Example 7

$1\frac{4}{7}\times3\frac{3}{4}\div\frac{30}{48}$

▶ ***Solution:*** $1\frac{4}{7}\times3\frac{3}{4}\div\frac{30}{48}=\frac{11}{7}\times\frac{15}{4}\times\frac{48}{30}=\frac{11}{7}\times\frac{15}{4}\times\frac{24}{15}=\frac{11}{7}\times\frac{1}{1}\times\frac{6}{1}=\frac{66}{7}=9\frac{3}{7}$

Example 8

$5\frac{7}{11}\div5\frac{13}{16}\times8\frac{1}{4}$

▶ ***Solution:*** $5\frac{7}{11}\div5\frac{13}{16}\times8\frac{1}{4}=\frac{62}{11}\div\frac{93}{16}\times\frac{33}{4}=\frac{62}{11}\times\frac{16}{93}\times\frac{33}{4}=\frac{2\times4\times3}{1\times3\times1}=8$

Practice

■Level A

1. Simplify the multiplication, and then find the answer.

a) $\frac{9}{99}\times\frac{7}{49}$ **b)** $\frac{25}{75}\times\frac{35}{105}$ **c)** $\frac{21}{77}\times\frac{11}{63}$

d) $\frac{36}{40}\times\frac{45}{54}$ **e)** $\frac{14}{49}\times\frac{13}{39}$ **f)** $\frac{49}{121}\times\frac{11}{56}$

g) $\frac{24}{42}\times\frac{7}{56}$ **h)** $\frac{40}{196}\times\frac{14}{20}$ **i)** $\frac{14}{144}\times\frac{12}{35}$

2. Solve these multiplication problems.

a) $3\frac{1}{2}\times4\frac{3}{5}$ **b)** $2\frac{7}{10}\times2\frac{1}{2}$ **c)** $3\frac{2}{5}\times3\frac{1}{2}$

d) $4\frac{1}{2}\times 3\frac{1}{4}$ **e)** $3\frac{1}{3}\times 4\frac{4}{5}$ **f)** $3\frac{5}{6}\times 4\frac{2}{5}$

g) $5\frac{5}{9}\times 4\frac{6}{15}$ **h)** $3\frac{5}{7}\times 5\frac{1}{4}$ **i)** $2\frac{5}{8}\times 5\frac{5}{7}$

■Level B

1. Solve these multiplication problems.

a) $\frac{3}{13}\times\frac{117}{39}$ **b)** $\frac{14}{196}\times\frac{28}{84}$ **c)** $\frac{22}{33}\times\frac{25}{75}$

d) $\frac{23}{196}\times\frac{14}{69}$ **e)** $\frac{21}{63}\times\frac{77}{154}$ **f)** $\frac{12}{36}\times\frac{13}{39}$

g) $\frac{21}{84}\times\frac{77}{154}$ **h)** $\frac{121}{33}\times\frac{39}{169}$ **i)** $\frac{121}{44}\times\frac{26}{169}$

2. A jug contains enough water to fill 2 glasses when the jug is $\frac{1}{4}$ full.

a) What fraction of the jug is needed to fill 5 glasses?

b) If the full jug contains 950 mL water, how much water does one glass hold?

3. 183 students went on a field trip. 5 buses were filled and 8 students traveled in cars. How many students were in each bus?

4. Solve the division problems.

a) $\frac{3}{9} \div \frac{18}{36}$ **b)** $\frac{11}{99} \div \frac{72}{81}$ **c)** $\frac{3}{16} \div \frac{18}{36}$

d) $\frac{12}{144} \div \frac{75}{25}$ **e)** $\frac{48}{196} \div \frac{24}{96}$ **f)** $\frac{12}{99} \div \frac{72}{81}$

g) $\frac{16}{112} \div \frac{45}{15}$ **h)** $\frac{14}{19} \div \frac{24}{76}$ **i)** $\frac{40}{196} \div \frac{20}{28}$

5. Solve the division problems.

a) $3\frac{1}{2} \div 4\frac{3}{4}$ **b)** $4\frac{6}{9} \div 4\frac{1}{4}$ **c)** $3\frac{7}{8} \div 5\frac{1}{4}$

d) $3\frac{1}{4} \div 3\frac{3}{8}$ **e)** $4\frac{2}{5} \div 2\frac{2}{5}$ **f)** $3\frac{1}{2} \div 2\frac{3}{4}$

g) $2\frac{1}{7} \div 3\frac{1}{8}$ **h**) $2\frac{1}{3} \div 4\frac{1}{2}$ **i**) $3\frac{2}{7} \div 4\frac{1}{3}$

6. The sum of two numbers is 70. The sum of $\frac{1}{8}$ of one number and $\frac{1}{3}$ of the other number is 15. Find these two numbers.

7. Robert collected 132 conkers but lost $\frac{1}{6}$ of them on his way to school through a hole in his bag. When he arrived at school, he gave $\frac{1}{5}$ of the remaining conkers to his friends. How many conkers did Robert leave in the end?

8. A truck traveled $100\frac{3}{4}$ km in $1\frac{5}{8}$ hrs. What was the truck's average speed in km/h?

■Level C

1. Rose grows blueberries on a farm and opens it to tourists for picking during the summer. One day a group of tourists came and picked $\frac{1}{3}$ of the berries. The next day another smaller group visited the farm and picked as many as the previous group. On the third day, a bus load of foreign tourists came and picked $\frac{1}{4}$ of the remaining berries. On the fourth day a storm hit the farm, and Rose lost half of what was left from the third day. If Rose had sold all the berries to a local supermarket before tourists came, she would have made \$70 000. How much will Rose make if she sells all the remaining berries after the storm? Round the answer to the nearest dollar.

2. In a lottery called Lotto 538, 5 unique numbers from 1 to 38 are drawn, and a ticket that matches all 5 numbers wins the jackpot. Nick bought 150 tickets, Steven bought the same number of tickets as Nick, and Tony bought $\frac{9}{2}$ as many tickets as Nick and Steven combined.

a) What is their probability of winning the jackpot if they combine all their tickets? Show the exact value in your answer.

b) Suppose the jackpot is \$50 000, and they agree to share the winnings equally. What is Tony's net profit, after subtracting the cost of tickets from the winnings? Assume each ticket costs \$2.50. Round the answer to the nearest dollar.

4.2 Multiplication and Division of Integers

When multiplying and dividing integers, there are 2 parts to consider: the **Value** and the **Sign**.

The Value of Integers

The procedure for calculating the value of the product or quotient of integers is the same as that of whole numbers.

The Sign of products

1. Determine how many negative numbers are in the question.
2. If the number of negative numbers is an odd number, the final integer is negative. Otherwise, the final integer is positive.

Example

Calculate the following products.

a. $(-3)\times(+2)$ **b.** $(-6)\div(-3)$ **c.** $(-4)\times(+2)\times(+3)$ **d.** $(-8)\div(+2)\times 3$

▶ *Solution*:

a. $(-3)\times(+2)=-(3\times 2)=-6$

The quantity of negative numbers is 1 (an odd number), so the final answer is negative.

b. $(-6)\div(-3)=+(6\div 3)=+2$

The quantity of negative numbers is 2 (an even number), so the final answer is positive.

c. $(-4)\times(+2)\times(+3)=-(4\times 2\times 3)=-24$

The quantity of negative numbers is 1 (an odd number), so the final answer is negative.

d. $(-8)\div(+2)\times 3=-(8\div 2\times 3)=-12$

The quantity of negative numbers is 1 (an odd number), so the final answer is negative.

The numbers "3" and "+3" have the same meaning.

Practice

■Level A

1. Calculate.

a) $(+4)\times(+3)$ **b)** $(+2)\times(+9)$ **c)** $(+3)\times(+7)$

d) $(+4)\times(+3)$ **e)** $(+3)\times(-5)$ **f)** $(+4)\times(-6)$

g) $(+7)\times(-4)$ **h)** $(+2)\times(-8)$ **i)** $(-4)\times(+6)$

j) $(-3)\times(+8)$ **k)** $(-8)\times(+3)$ **l)** $(-5)\times(+4)$

m) $8\times(-2)$ **n)** $6\times(-5)$ **o)** $10\times(-4)$

p) $9\times(-3)$ **q)** 5×4 **r)** 6×9

s) $6\times(-7)$ **t)** -4×8 **u)** $(-5)\times(-4)$

2. Calculate.

a) $(+6)\div(+3)$ **b)** $(+9)\div(+3)$ **c)** $(+10)\div(+2)$

d) $(+8)\div(+4)$ **e)** $(+10)\div(-5)$ **f)** $(+8)\div(-2)$

g) $(+9)\div(-3)$ **h)** $(+8)\div(-2)$ **i)** $(-12)\div(+6)$

j) $(-14)\div(+7)$ **k)** $(-24)\div(+6)$ **l)** $(-25)\div(+5)$

m) $8\div(-2)$ **n)** $12\div(-4)$ **o)** $(-14)\div(-7)$

p) $18\div(-6)$ **q)** $(-12)\div3$ **r)** $(-10)\div(-5)$

s) $(-12)\div(-6)$ **t)** $(-18)\div3$ **u)** $(-16)\div(-2)$

■Level B

1. Calculate.

a) $(+42)\times(+13)$ **b)** $(+23)\times(+25)$ **c)** $(+42)\times(+17)$

d) $(+33)\times(+15)$ **e)** $(+23)\times(-51)$ **f)** $(+26)\times(-31)$

g) $(+42)\times(-21)$ **h)** $(+61)\times(-21)$ **i)** $(-14)\times(+62)$

j) $(-23)\times(+42)\times(+2)$ **k)** $(-34)\times(+14)\times(+12)$

l) $38\times(-26)\times(-6)$ **m)** $28\times(-36)\times(+22)$

n) $25\times(-43)\div(-5)$ **o)** $51\times43\div(-17)$

p) $(-29)\times44\div(+11)$ **q)** $19\times(-36)\div(-9)$

2. Calculate.

a) $(+63)\div(+3)$ **b)** $(+96)\div(+6)$ **c)** $(+128)\div(+4)$

d) $(+64)\div(+4)$ **e)** $(+105)\div(-5)$ **f)** $(+135)\div(-9)$

g) $(+256)\div(-8)$ **h)** $(+105)\div(-7)$ **i)** $(-48)\div(+16)$

j) $(-156)\div(+13)\div(-4)$ **k)** $(-323)\div(+19)\div(-17)$

l) $84\div(-21)\div(-2)$ **m)** $384\div(-16)\div(8)$

n) $432\div(-27)\times(-6)$ **o)** $196\div14\times(-2)$

p) $-391\div(-23)\times(-11)$ **q)** $-448\div(-32)\times9$

3. Mary is 164 cm tall and her roommate Aileen is $\frac{7}{8}$ as tall as her. How tall is Aileen?

4. The weather forecaster says that it is 25℃ in Toronto. It is only $\frac{9}{10}$ as hot in Vancouver. What is the temperature in Vancouver?

5. There are 28 pupils in the class. $\frac{4}{7}$ of the pupils support Tony for class president and the remaining people support Daniel. How many pupils support Daniel?

6. The local shop normally sells T-shirts for \$25 each. The shopkeeper says that students at the secondary school can buy the T-shirts for $\frac{1}{5}$ less than the normal price. How much can the students at the secondary school buy the T-shirts for?

■Level C

1. In a particular course, the passing mark is 65. Johnny's mark is 5% higher than the passing mark. Chris' mark is 15% higher than Johnny's, and Eric's mark is 10% lower than the average mark of Johnny and Chris. Did Eric pass the course? What was Eric's mark?

2. A snail climbs up a wall that is 15 m in height. On Day 1 it climbs 3 m, on Day 2 it falls by 2 m, and on Day 3 it climbs another 1 m. Assume on Day 4 the snail will continue with the pattern as Day 1, Day 5 as Day 2, and so on. How many days will it take for the snail to reach the top of the wall?

4.3 Order of Operations

The Order of Operations—BEDMAS

The order of the operations follows the rule of **BEDMAS.**

Priority Level	Level 1	Level 2	Level 3	Level 4
Operation Name and First Letter	**B** (brackets)	**E** (exponents)	**D** (division) **M** (multiplication)	**A** (addition) **S** (subtraction)

1. The order of operations in a problem starts with Level 1, and goes to Level 2, Level 3 and Level 4 from left to right.
2. If the BEDMAS rule is not followed exactly, your answer will be incorrect.

Example 1

Write out the priority level of the following operations.

$100-45\div15(10-3^2)+10$

▶ *Solution*:

```
100-45 ÷ 15(10-3²)+10
                └┘
                1
      └──┘    └─┘
       2       2
       └───────┘
           3
   └──────┘
      4
       └──────────────┘
             5
```

Example 2

Calculate the question step by step based on the priority level.

$$\frac{4^2\times(99-(25+44)-10)}{36\times27\div9}$$

▶ *Solution*:

$$\frac{4^2\times(99-(25+44)-10)}{36\times27\div9}=\frac{16\times(99-69-10)}{36\times27\div9}$$

$$=\frac{16\times20\times9}{36\times27}=\frac{4\times20\times1}{9\times3}=\frac{80}{27}=2\frac{26}{27}$$

Operations on Fractions and Integers

Example 3

Simplify the following.

$$\frac{(-\frac{1}{3})^2\times(10-((-1\frac{1}{2})-(-1\frac{1}{3}))-(\frac{2}{3}))}{(-\frac{1}{3})\times(-\frac{1}{3})\div\frac{1}{6}}$$

▶ *Solution*:

$$\frac{(-\frac{1}{3})^2\times(10-((-1\frac{1}{2})-(-1\frac{1}{3}))-(\frac{2}{3}))}{(-\frac{1}{3})\times(-\frac{1}{3})\div\frac{1}{6}}=\frac{(\frac{1}{9})\times(10-((-\frac{3}{2})+(\frac{4}{3}))-(\frac{2}{3}))}{(\frac{1}{3})\times(\frac{1}{3})\times\frac{6}{1}}$$

$$=\frac{(\frac{1}{9})\times(10-(-\frac{1}{6})-(\frac{2}{3}))}{\frac{2}{3}}=(\frac{1}{9})\times(10+\frac{1}{6}-\frac{2}{3})\times\frac{3}{2}=(\frac{1}{9})\times(\frac{60+1-4}{6})\times\frac{3}{2}$$

$$=\frac{1}{9}\times\frac{57}{6}\times\frac{3}{2}=\frac{1}{3}\times\frac{19}{2}\times\frac{1}{2}=\frac{19}{12}=1\frac{7}{12}$$

Practice

■Level A

Calculate.

a) $1\frac{1}{3}\times2\frac{1}{2}$

b) $2\frac{1}{4}\times2\frac{1}{3}$

c) $1\frac{3}{4}\times2\frac{2}{3}$

d) $2\frac{2}{5}\times3\frac{1}{3}$

e) $1\frac{1}{4}\times2\frac{1}{8}$

f) $1\frac{1}{4}\times2\frac{2}{15}$

g) $3\frac{1}{2}\times3\frac{1}{7}$ **h)** $2\frac{1}{12}\times2\frac{2}{15}$ **i)** $1\frac{1}{17}\times2\frac{5}{6}$

j) $(-1\frac{1}{9})\times(+5\frac{2}{5})$ **k)** $(-2\frac{1}{13}\times(+1\frac{4}{9})$ **l)** $(-2\frac{3}{16})\times(+1\frac{11}{15})$

m) $(-3\frac{3}{11}\times(+3\frac{1}{18}))$ **n)** $(-2\frac{2}{3})\times(-5\frac{1}{4})$ **o)** $(-1\frac{1}{12})\times(-3\frac{1}{13})$

■Level B

1. Calculate.

a) $1\frac{1}{2}\div\frac{3}{8}$ **b)** $1\frac{1}{3}\div2\frac{2}{7}$ **c)** $1\frac{1}{4}\div3\frac{3}{4}$

d) $2\frac{1}{5}\div1\frac{7}{15}$ **e)** $(-1\frac{1}{2})\div(-\frac{5}{18})$ **f)** $(-2\frac{1}{3})\div(-3\frac{8}{9})$

g) $(-1\frac{1}{4})\div(-1\frac{9}{16})$ **h)** $(-1\frac{1}{5})\div(-2\frac{2}{5})$ **i)** $(-1\frac{3}{5})\div(+3\frac{1}{5})$

j) $(-2\frac{1}{3})\div(+1\frac{5}{9})$ **k)** $(-2\frac{1}{4})\div(+8\frac{1}{2})$ **l)** $-1\frac{1}{4}\div2\frac{1}{4}$

m) $2\frac{2}{7}\div(-1\frac{2}{7})$ **n)** $+1\frac{2}{3}\div(-4\frac{3}{8})$ **o)** $1\frac{1}{5}\div(-1\frac{1}{2})$

2. Calculate.

a) $(+2\frac{2}{7})\times(-\frac{7}{16})\times(-1\frac{1}{2})$ **b)** $(1\frac{2}{3})\times(-\frac{18}{25})\times(-2\frac{1}{2})$

c) $(-1\frac{2}{3})\times(\frac{25}{36})\times(-1\frac{3}{20})$ **d)** $(-2\frac{2}{5})\times(\frac{7}{12})\times(-2\frac{2}{7})$

e) $(-2\frac{3}{5})\times(-\frac{15}{32})\times(-1\frac{1}{6})$

f) $(-1\frac{5}{6})\times(-2\frac{6}{15})\times(-2\frac{1}{4})$

g) $(-2\frac{1}{2})\times(+\frac{1}{4})\div(+1\frac{3}{4})$

h) $(-2\frac{1}{4})\times(2\frac{7}{12})\times(-2\frac{1}{7})$

i) $(-1\frac{6}{15})\div(+1\frac{24}{25})\times(+2\frac{15}{20})$

j) $(-2\frac{1}{4})\times(2\frac{7}{12})\times(-1\frac{11}{14})$

3. Simplify the following.

a) $\dfrac{((-2\frac{1}{4})-(-3\frac{1}{3})-(1\frac{1}{2})(-2\frac{1}{4}))\times(2\frac{7}{12})}{(-\frac{3}{4})\div(-3\frac{1}{3})}$

b) $\dfrac{(-1\frac{2}{3})+(-2\frac{5}{6})-(2\frac{2}{3})}{(-\frac{1}{3})\div(-\frac{1}{3})}$

c) $\dfrac{(+1\frac{2}{3})\times(-2\frac{4}{6})\div(2\frac{2}{6})}{(-\frac{1}{2})\div(-\frac{2}{3})\times(1\frac{8}{12})}$

Level C

1. In a family, Abraham is Homer's father, and Homer is Bart's father. Bart is 15 years old now. Five years ago, Bart was $\frac{1}{4}$ as old as Abraham. In 15 years, Homer will be 1.5 times as old as Bart. What is the age difference between Abraham and Homer now?

2. Joseph bought 2 pizzas one day: one Hawaiian and one Greek. He ate $\frac{3}{8}$ of the Hawaiian for dinner on Day 1. On Day 2 he had friends over and they had $\frac{1}{2}$ of what was left of the Hawaiian, and $\frac{2}{3}$ of the Greek. On Day 3 Joseph took out whichever pizza had a bigger leftover from Day 2 and ate another $\frac{1}{2}$ of it. How much of each pizza was left at the end of Day 3?

Chapter 4 TEST

1. Simplify and calculate.

a) $\frac{9}{10}\times\frac{5}{6}$ **b)** $\frac{2}{7}\times\frac{13}{39}$ **c)** $\frac{7}{121}\times\frac{11}{8}$

d) $\frac{24}{7}\times\frac{7}{56}$ **e)** $\frac{4}{196}\times\frac{14}{2}$ **f)** $\frac{24}{144}\times\frac{6}{36}$

2. Simplify and calculate.

a) $3\frac{1}{3}\times3\frac{1}{2}$ **b)** $3\frac{3}{4}\times2\frac{2}{5}$ **c)** $4\frac{1}{6}\times3\frac{3}{5}$

d) $4\frac{2}{3}\times4\frac{1}{8}$ **e)** $4\frac{1}{8}\times5\frac{1}{11}$ **f)** $3\frac{3}{8}\times5\frac{1}{3}$

3. Simplify and calculate.

a) $\frac{13}{225}\times\frac{15}{39}$ **b)** $\frac{36}{49}\times\frac{35}{54}$ **c)** $\frac{12}{48}\times\frac{14}{84}$

d) $\frac{21}{63}\times\frac{77}{154}$ **e)** $\frac{121}{33}\times\frac{39}{169}$ **f)** $\frac{22}{55}\times\frac{68}{51}$

4. Simplify and calculate.

a) $\frac{13}{196}\div\frac{75}{25}$ **b)** $\frac{24}{98}\div\frac{48}{192}$ **c)** $\frac{12}{99}\div\frac{27}{72}$

d) $\frac{18}{114}\div\frac{45}{15}$ **e)** $\frac{14}{19}\div\frac{12}{76}$ **f)** $\frac{44}{196}\div\frac{20}{28}$

5. Calculate.

a) $35\times(-23)$ **b)** 31×46 **c)** $+16\times(+28)$

d) $(-19)\times34$ **e)** $29\times(-27)$ **f)** $(-42)\times(-37)$

6. Calculate.

a) $(+288)\div(-8)$ **b)** $(+324)\div(-9)$ **c)** $(-351)\div(+13)$

d) $(-560)\div(+16)$ **e)** $(-646)\div(+34)$ **f)** $(-837)\div(+31)$

7. Simplify and calculate.

a) $(-1\frac{1}{8})\times(+6\frac{2}{5})$ **b)** $(-3\frac{1}{11})\times(+1\frac{4}{17})$ **c)** $(-1\frac{3}{12})\times(+2\frac{6}{15})$

d) $(2\frac{1}{12})\times(-2\frac{2}{15})$ **e)** $(-3\frac{1}{4})\times(-2\frac{2}{13})$ **f)** $(-6\frac{2}{3})\times(-3\frac{1}{10})$

8. Simplify and calculate.

a) $(-3\frac{1}{3})\div(+4\frac{2}{7})$ **b)** $(-3\frac{1}{2})\div(-\frac{35}{26})$ **c)** $(-2\frac{1}{4})\div(1\frac{11}{16})$

d) $(2\frac{2}{7})\div(-1\frac{11}{21})$ **e)** $(-2\frac{2}{3})\div(-4\frac{3}{8})$ **f)** $(-2\frac{1}{5})\div(+2\frac{18}{35})$

9. Simplify and calculate.

a) $1\frac{3}{4}\times\frac{25}{21}\times1\frac{7}{20}$ **b)** $(-3\frac{5}{6})\times(\frac{7}{23})\times(-2\frac{2}{7})$

c) $(\frac{1}{9})\times(-1\frac{1}{7})$ **d)** $(-2\frac{3}{4})\times(2\frac{5}{11})\times(-2\frac{2}{3})$

10. Simplify and calculate.

a) $\dfrac{(1\frac{1}{6}+(-4\frac{1}{3})+(3\frac{1}{2})(-5\frac{1}{3}))\times4\frac{1}{2}}{(-4\frac{3}{4})\div(6\frac{1}{3})}$

b) $\dfrac{-(1\frac{2}{3})^2+(-2\frac{5}{6})-(-1\frac{1}{3})^2}{(-\frac{1}{3})^2+(-\frac{2}{3})^2}$

c) $\dfrac{((3\frac{3}{4})(-3\frac{1}{3})-(2\frac{1}{2})(-2\frac{2}{3}))\div(1\frac{1}{24})}{(-\frac{3}{4})+(-3\frac{1}{3})-(-4\frac{5}{6})}$

11. The weather forecaster says that it is 30℃ in Toronto. In Vancouver it is only $\frac{4}{5}$ as hot as Toronto. What is the temperature in Vancouver?

12. A grocery store normally sells rice for \$45 each bag. The storekeeper says that on Canada Day, the sale price of a bag of rice can be $\frac{1}{9}$ less than the normal price on Canada Day. How much is the sale price of a bag of rice on Canada Day?

13. A jug contains enough tea to fill 3 cups when the jug is $\frac{1}{4}$ full.

a) How full is the jug if the jug is expected to fill 4 cups?

b) If the full jug contains 3600 mL tea, how much tea does one cup hold?

14. The sum of two numbers is 80. The sum of $\frac{1}{5}$ of the first number and $\frac{1}{3}$ of the other number is 15. Find these two numbers.

15. A school bus traveled $110\frac{1}{2}$ km in $1\frac{5}{8}$ hrs. What was the school bus' average speed in km/h?

16. Daniel decided to start jogging to stay in shape. The following are the distances he jogged last week:

Monday: $\frac{3}{2}$ km

Tuesday: $\frac{3}{5}$ km

Wednesday: $\frac{1}{2}$ of the distance on Monday

Thursday: $\frac{2}{3}$ of the distance on Tuesday

Friday: average distance from Monday to Wednesday

Saturday: average distance from Tuesday to Thursday

Sunday: average distance from Monday to Saturday

What is the total distance that Daniel jogged last week? Give the answer as an exact value.

CHAPTER 5

Variables and Relations

5. 1 Variables

5. 2 Three Ways to Describe Relations between Two Variables

5. 3 Linear Relations

Chapter 5 TEST

Chapter 5 Variables and Relations

5.1 Variables

Independent Variables and Dependent Variables

In a system where an input causes a change in the output, the input is called the **independent variable**, and the output is called the **dependent variable.** Independent variables are commonly represented by x, and dependent variables are commonly represented by y.

When such a system is described using a table of values, we usually write the independent variables in the left column or the top row, and the dependent variables in the right column or the bottom row. When the system is drawn as a graph, independent variables are plotted on the x-axis, and dependent variables are plotted on the y-axis.

Ordered pairs (x, y) can be used to describe a point on a coordinate plane. When there are two or more points, we can draw a line that describes the system.

Example 1

Consider the following table for distance driven by a car. Which variable is independent, and which one is dependent?

Time (hr)	1	2	3	4	5
Distance (km)	70	140	210	280	350

▶ *Solution*:

The input is the time in hours, and the output is the distance in kilometres. Therefore, time is the independent variable, and distance is the dependent variable.

Example 2

Decide which variable is independent, and which one is dependent.

Time Elapsed (hr)	1	2	3	4	5
Battery Remaining (%)	85	70	55	40	25

▶ *Solution*:

Since the battery level changes with respect to time, time is the independent variable, and battery level is the dependent variable.

Relations, Elements, Domain and Range

A set of ordered pairs (x, y) is also called **relation**, and the values of x and y are referred to as **elements**. The set of all values the first element in the ordered pair can take is called the **domain**, and that of the second element is called the **range**.

Example 3

Set up the table and draw a graph on a coordinate plane based on the ordered pairs of coordinates (x, y).

$A(-2, -1)$, $B(-1, 0)$, $C(1, 2)$, $D(3, 4)$

▶ *Solution*:

a) Set up the table

Value \ Point	A	B	C	D
x	-2	-1	1	3
y	-1	0	2	4

b) Draw the graph

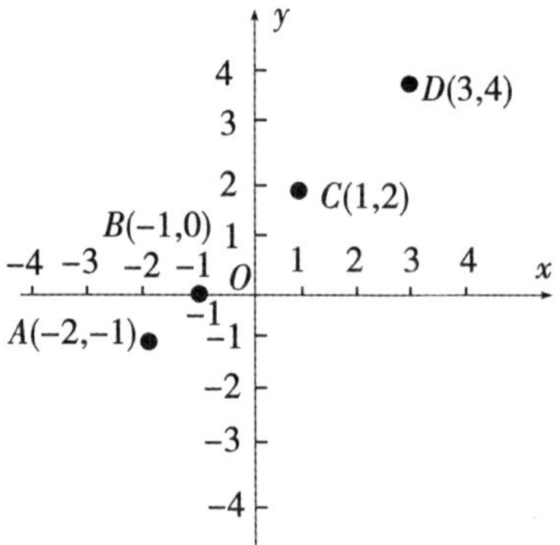

Values of Variable Expressions

When given values of variables, you can use them to calculate the value of an expression.

Example 4

Assume $a=2$ and $b=-3$, simplify the expression, $2a-ab+b^2$.

▶ *Solution*:

$2\times2-2\times(-3)+(-3)^2=4-(-6)+9=19$

Practice

Level A

1. Based on the coordinates of the following points, $A(-3, -4)$, $B(-4, -2)$, $C(-2, 0)$, $D(0, 2)$, $E(1, 2)$, $F(2, 3)$, $G(4, 4)$

a) Fill in the table.

Points / Value	A	B	C	D	E	F	G
x							
y							

b) Draw the points on the coordinate plane.

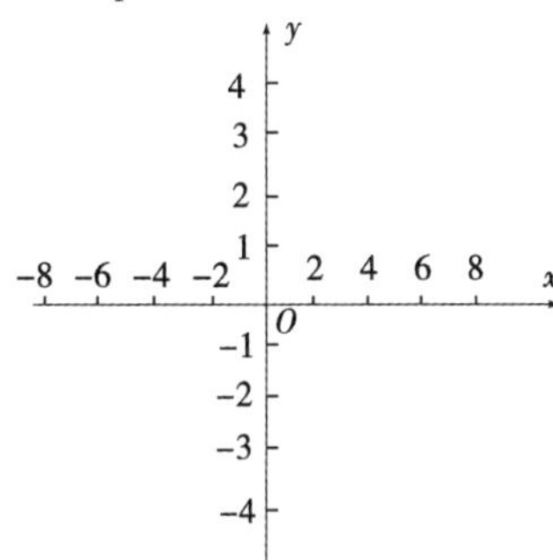

2. Change the following values to ordered pairs. Set up a table and plot the points on the coordinate plane.

a) $x=3$, $y=2$ b) $x=2$, $y=4$ c) $x=3$, $y=4$ d) $x=4$, $y=2$

e) $x=4$, $y=5$ f) $x=3$, $y=1$ g) $x=-3$, $y=2$ h) $x=2$, $y=-2$

i) $x=-2$, $y=-5$ j) $x=-4$, $y=2$ k) $x=-3$, $y=-2$ l) $x=3$, $y=-2$

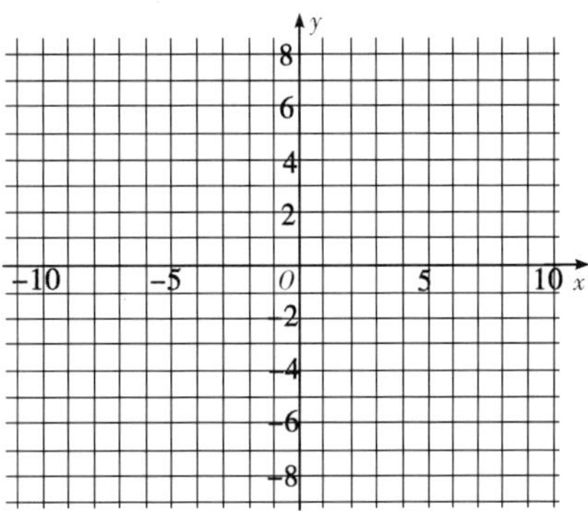

Level B

1. Assume $a=2$, $b=-3$ and $c=-2$. Simplify the following expressions.

a) $3a-ab+b^2$

b) $3a-2ab-3b$

c) $3ab+4b-3ba$

d) $\dfrac{7ab\div 14}{6c+4b}$

e) $\dfrac{ab\times a}{6a-b}$

f) $\dfrac{5b\div 5a}{6c^2+4b^2}$

g) $\dfrac{7(2a+2b)\div 14}{3a\times 2c}$

h) $\dfrac{(3a-2b)\div 3}{2a\times 4c}$

i) $\dfrac{b(2a^2+2b^2)}{2ab-2ac}$

j) $\dfrac{b^2+c^2+2a^2}{2abc}$

k) $\dfrac{b^2+c^2-2a^2}{a^2bc^2}$

l) $\dfrac{ab(b^2+c-2a)}{a^2+b^2+c^2}$

2. Fill in the spaces in the following tables. Find the relations between the x values and y values, and draw graphs of the relations.

a)

x	2	3	4	5	6	7	8
y	5	6	7	8			

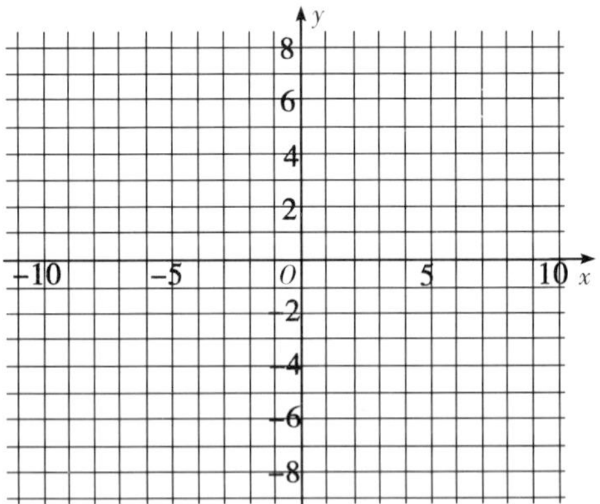

b)

x	2	3	4	5	6	7	8
y	8	7	6	5			

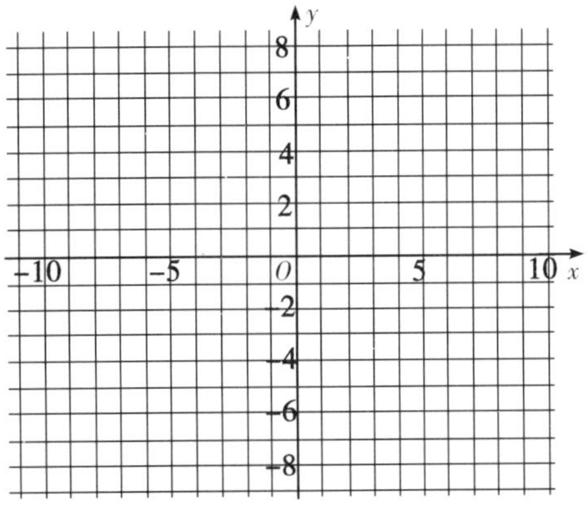

c)

x	2	3	4	5	6	7	8
y	6	6	6	6			

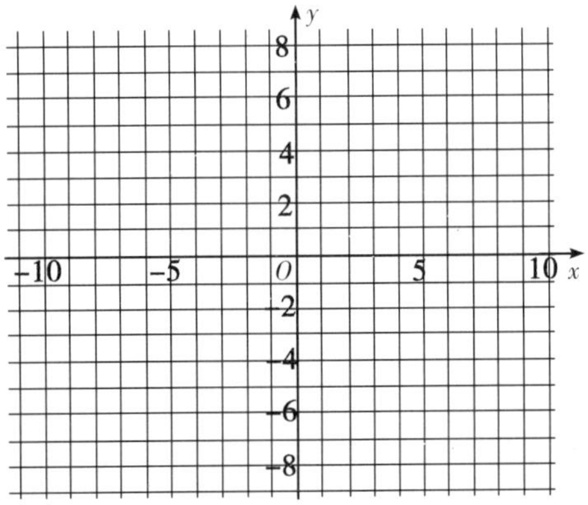

■Level C

1. At a carnival, Jim, Harry and Michael all work at a stand that sells mini-doughnuts. They receive a base pay, plus a bonus for each doughnut sold. Jim sold 400 doughnuts and made \$135. Harry sold 300 doughnuts and made \$105. Michael sold 700 doughnuts and made \$225.

a) How much bonus is paid for each doughnut sold?

b) What is the base pay?

c) If someone sold 550 doughnuts, how much would he earn?

2. At a fundraising event, cupcakes are sold to visitors. The event starts at 8 am. At 9 am, there are 250 cupcakes left; at noon there are 175 left; and at 3 pm there are 100 left.

a) On average, how many cupcakes are sold every hour?

b) How many cupcakes are there when the event starts?

c) When will the cupcakes be sold out?

5.2 Three Ways to Describe Relations between Two Variables

To describe the relations between two variables, we can use three ways: **formulas**, **tables** and **graphs**. Any one of the three can be transformed into the other two. In this book, we mainly focus on a linear relationship that has a straight line as the graph.

$y=mx+b$ is the basic formula to describe a linear relation. We can substitute the coordinates of any two points into the formula to find m and b.

Example

Set up the table and write the formula corresponding to the linear relation in the following graph.

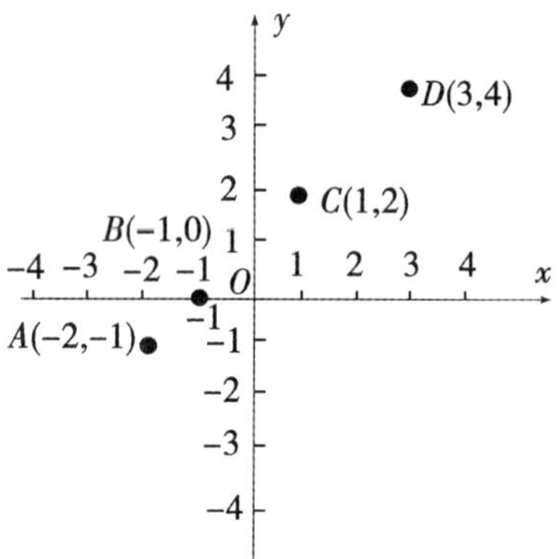

▶ *Solution*:

a) Set up the table as the following:

Points	Point A	Point B	Point C	Point D
x Value	-2	-1	1	3
y Value	-1	0	2	4

b) Set up a formula, and substitute $B(-1, 0)$ into $y=mx+b$,

So $0=m(-1)+b$, $b=m$ (1)

Substitute $C(1, 2)$ into $y=mx+b$, so $2=m\times 1+b$ (2)

Then, applying expression (1) to expression (2), we get $2=m+m=2m$, $m=1$, $b=m=1$. Since $m=1$ *and* $b=1$, the formula of the graph is $y=x+1$.

Practice

■Level A

1. Put these ordered pairs into tables and draw the corresponding graphs.

a) (3, 4), (4, 5), (5, 6), (6, 7)

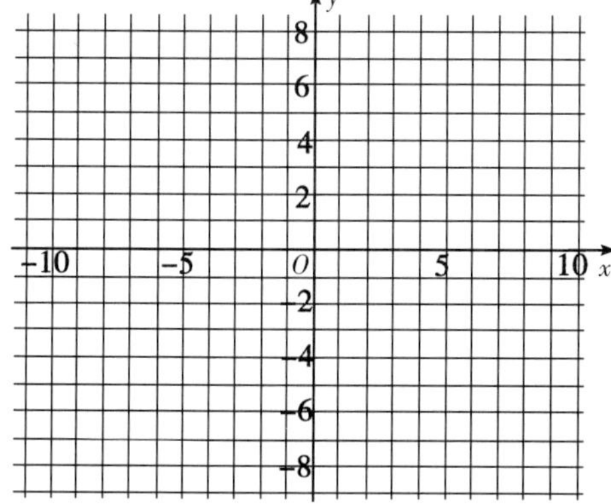

b) (−4, −6), (−3, −4), (−2, −2), (−1, 0)

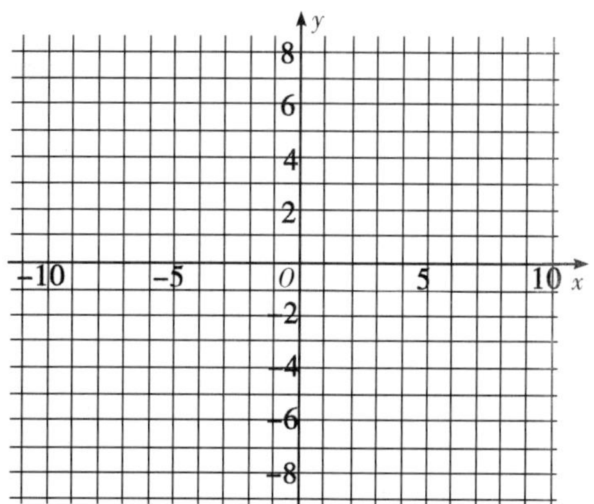

c) (1, 7), (2, 5), (3, 3), (4, 1)

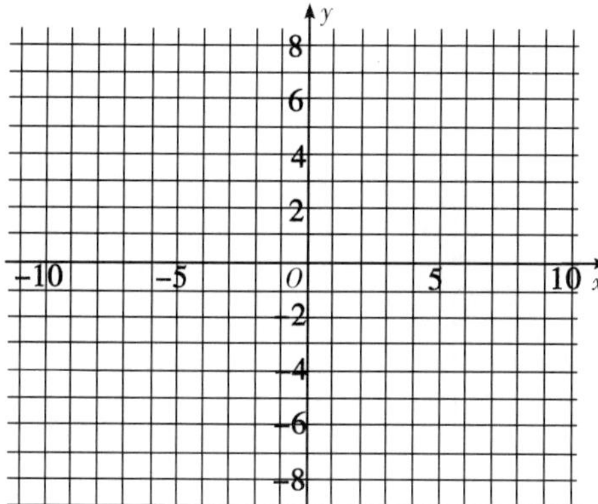

d) (−5, 5), (−4, 3), (−3, 1), (−2, −1)

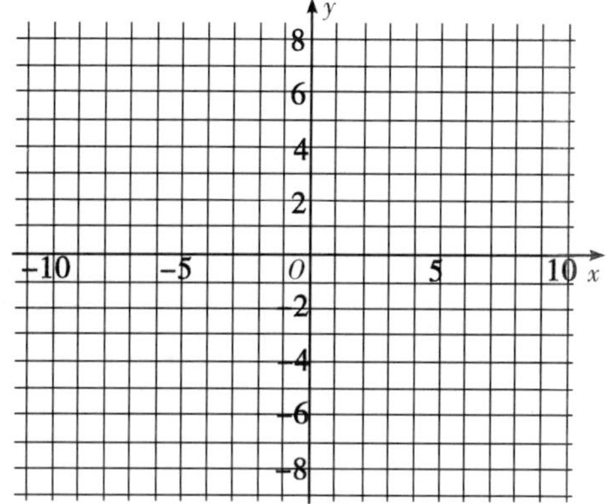

e) (−2, 3), (−1, 4), (0, 5), (1, 6)

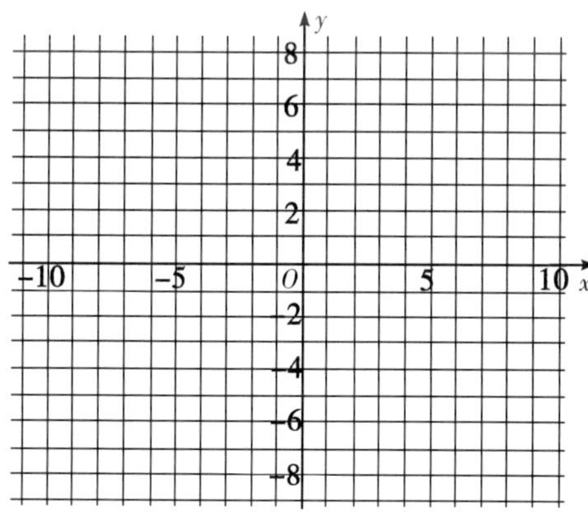

f) (0, 0), (1, 1), (2, 4), (3, 9)

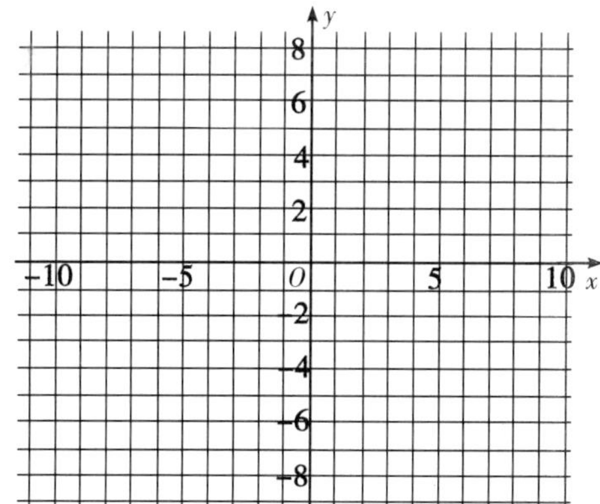

g) (−3, −9), (−2, −4), (−1, −1), (0, 0)

h) (−3, −6), (−2, −4), (−1, −2), (1, 2)

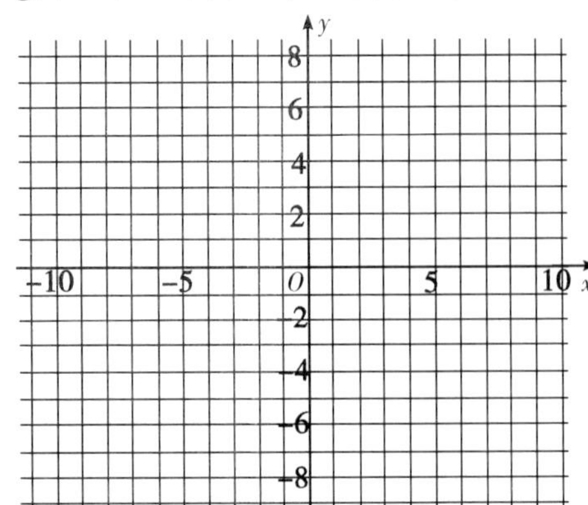

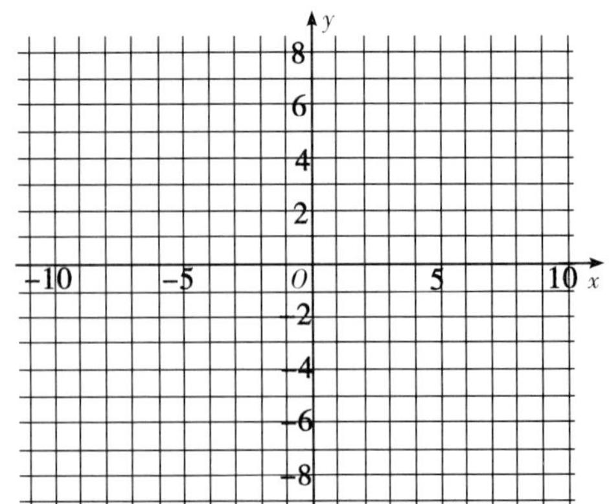

2. Complete the following tables and draw the graphs.

a)

x	−3	−2	−1	0	1	2
y	−2	0	2	4		

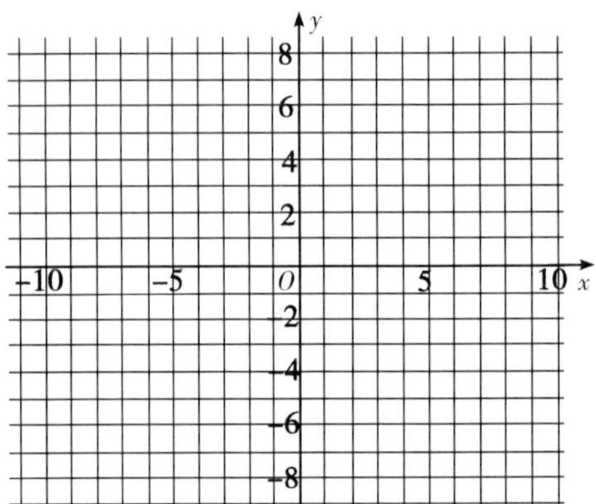

b)

x	−3	−2	−1	0	1	2	3
y	5	3	1	−1			

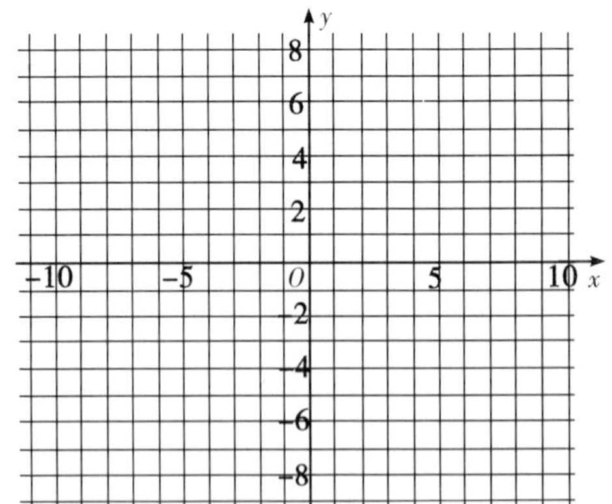

Level B

1. Based on the relations between x and y, draw the graphs.

a) $y=x$

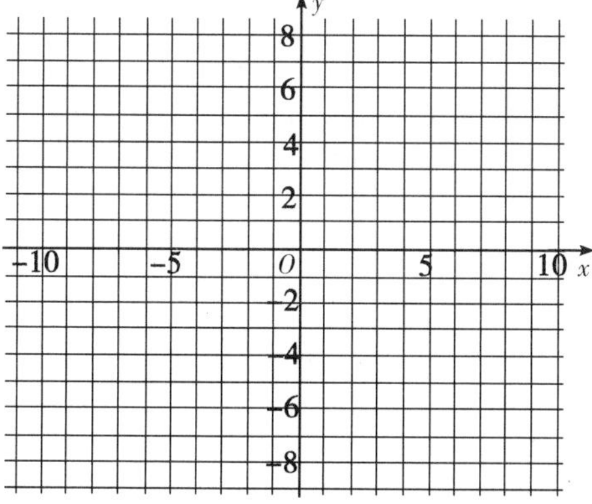

b) $y=2$

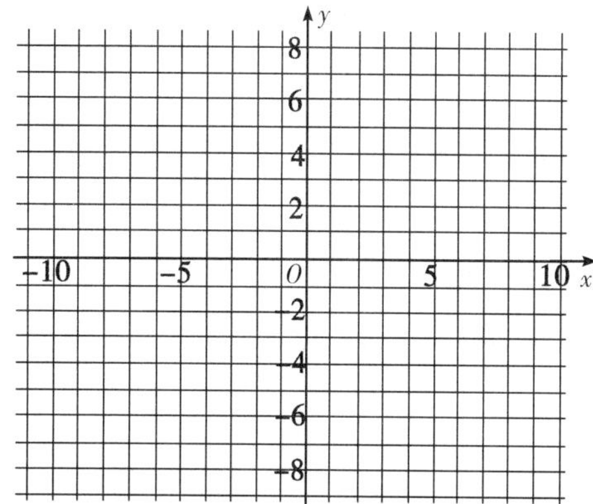

c) $y=3x-5$

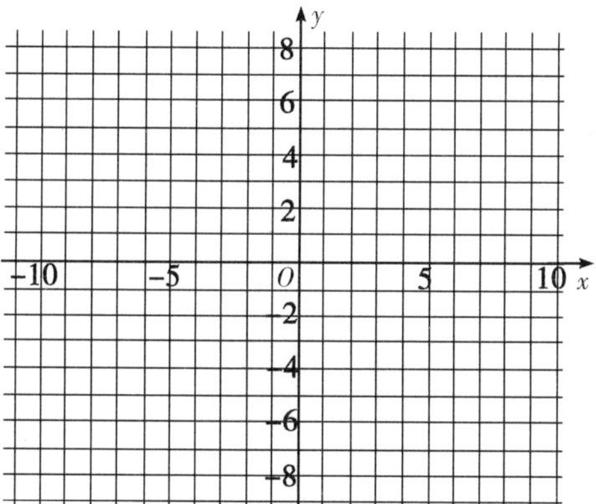

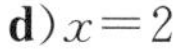

d) $x=2$

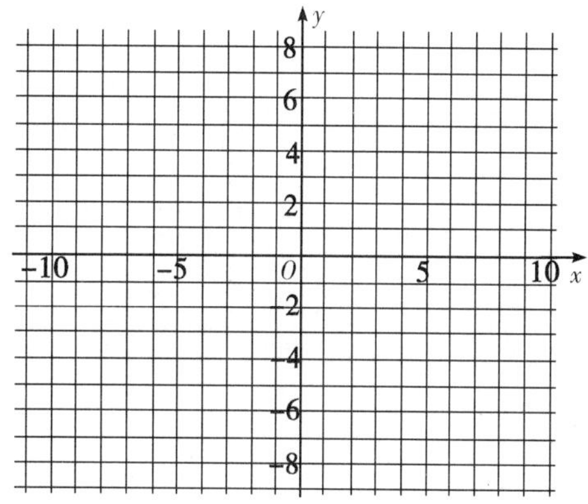

e) $y=x^2$

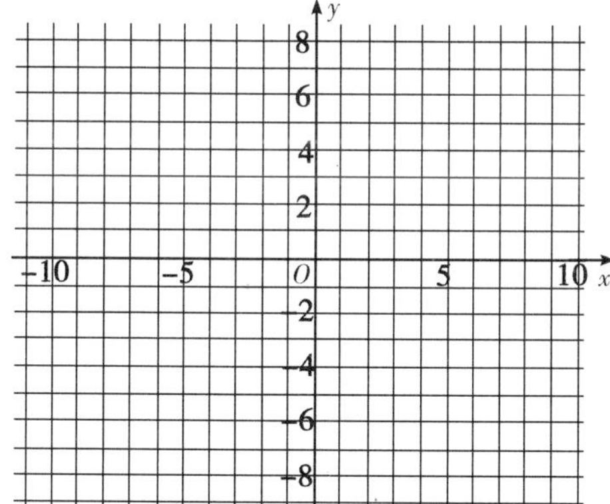

f) $y=2x^2$

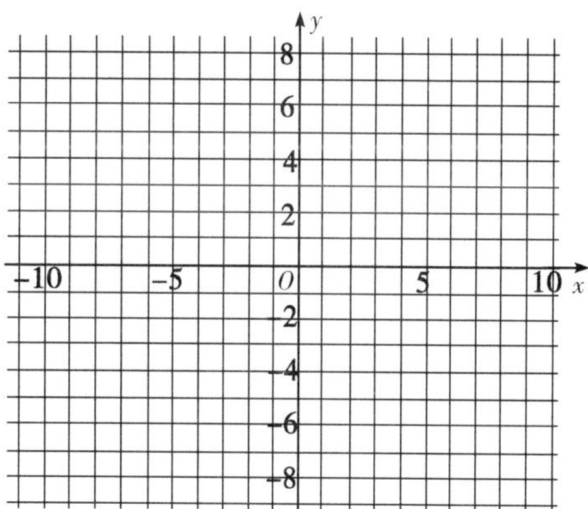

g) $y=x^2+1$

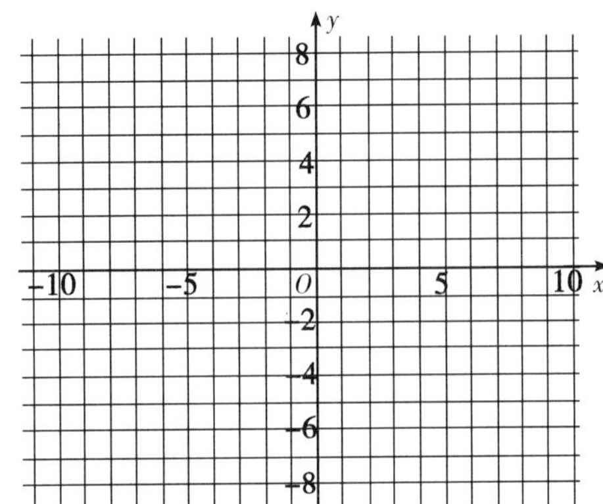

h) $y=x^2-1$

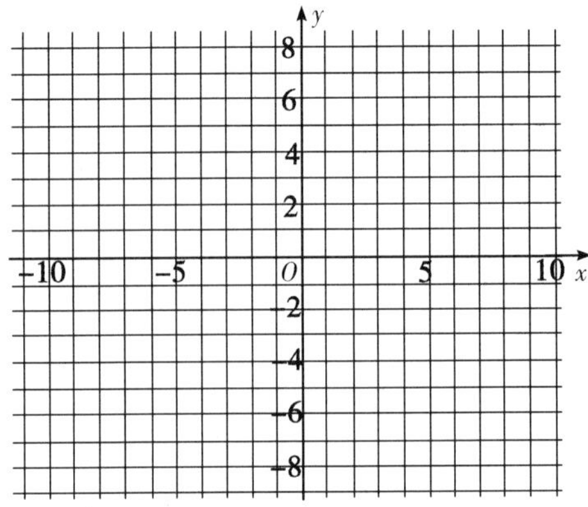

2. Complete the following tables and describe the relations between x and y with formulas.

a)

x	2	3	4	5	6	7
y	12	18	24	30		

b)

x	-2	-1	0	1	2	3
y	1	-2	-5	-8		

c)

x	-2	-1	0	1	2	3	4
y	3	1	-1	-3			

d)

x	0	1	2	3	4	5	6
y	0	1	4	9			

■Level C

1. State whether the following relations are linear, and write equations to describe the relations.

a) The area of a circle and its radius.

b) The tax paid in British Columbia when purchasing an item and the price of the item.

c) The circumference of a circle and its diameter.

d) The length of a square and its area.

e) The volume of a sphere and its radius.

2. A car travelling at 60 km/h needs 32.5 m to come to a complete stop. Assume the stopping distance varies directly with the square of its speed.

a) Determine the stopping distance for a car travelling at 120 km/h.

b) The stopping distance is 80 m. What is the speed of the car before it began to slow down?

5.3 Linear Relations

In a linear relations,

1. The graph is a straight line in the coordinate plane.

2. All of the coordinate pairs (x, y) that fit the equation will be on the line.

3. When the independent variable x increases by the value $\triangle x$, the dependent variable y increases by a value $\triangle y$. $\frac{\triangle y}{\triangle x}=m=$a constant. $\triangle$ means the amount of change so $\triangle x$ is the change in x and $\triangle y$ is the change in y.

m is called the slope of the line and it describes how steep the line is. $\triangle y$ is called the **Rise**, and $\triangle x$ is called the **Run**. So, $m=\frac{Rise}{Run}$.

Slopes can be negative or positive. Positive slopes signify that when x increases, y increases, while negative slopes signify that when x increases, y decreases.

Example 1

Which set of ordered coordinates is part of a linear relation?

a) $A(-2, 4)$, $B(-1, 2)$, $C(0, 0)$, $D(1, -2)$ **b)** $A(-2,-4)$, $B(-1,-2)$, $C(0,1)$, $D(1,2)$

▶ *Solution*:

When the value of x increases by 1, the corresponding y value decreases by 2, so (a) is part of a linear relation; In part (b), if you plot the coordinates on a graph, point c(0, 1) is not on the line, so (b) is not a linear relation.

Example 2

Based on the above coordinates in Example 1(a), draw a graph for the linear relation. Find the slope m and write the expression.

▶ *Solution*:

Draw the points first, and then connect all 4 points to form the line.

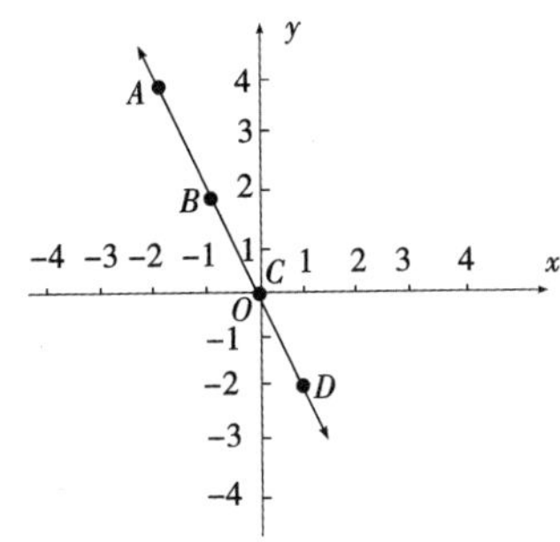

Slope $m=\frac{Rise}{Run}=\frac{-6}{3}=-2$

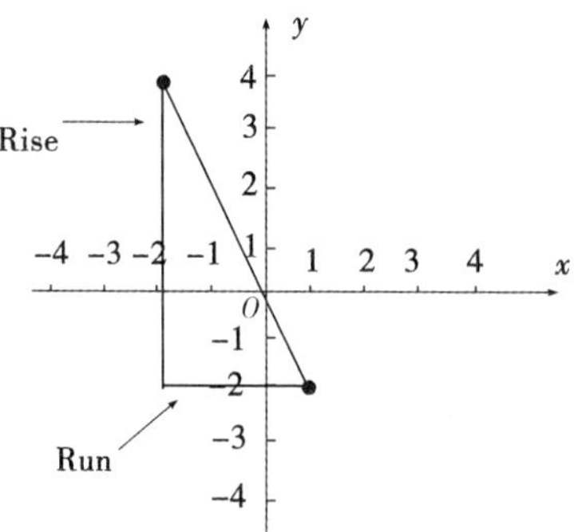

As a rule, if you try to connect the points of a linear relation and they do not fall exactly on the edge of a ruler, they are most likely plotted incorrectly. All points of a linear relation must form a straight line.

Practice

■Level A

Based on the coordinates, draw a graph and identify if it is a linear relation or not.

a) $A(-3,-4)$, $B(-2,-2)$, $C(-1,0)$, $D(0,2)$

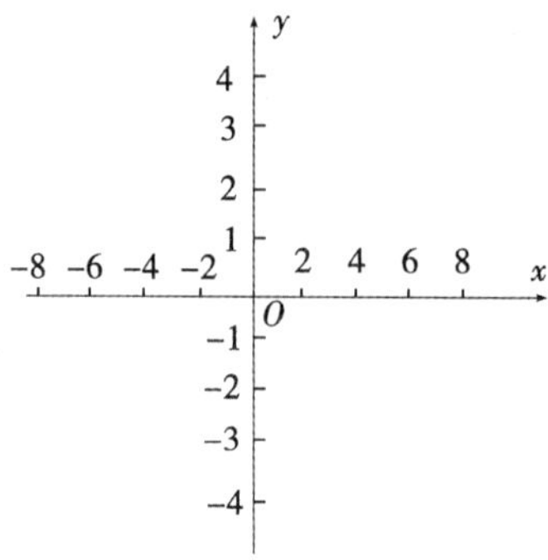

b) $A(0,5)$, $B(1,2)$, $C(2,0)$, $D(-3,-3)$

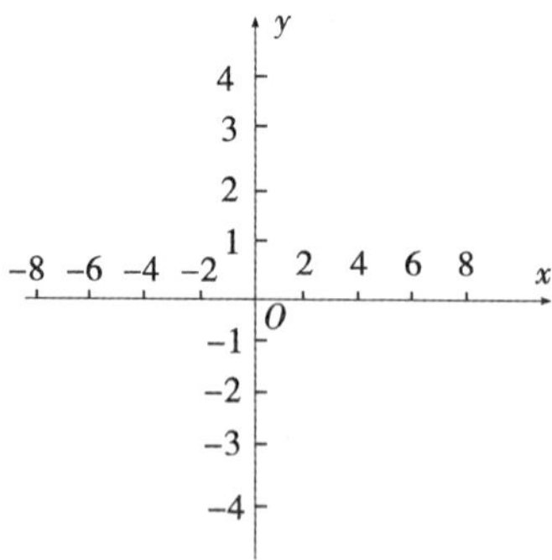

c) $A(0,0)$, $B(1,1)$, $C(2,4)$, $D(3,9)$

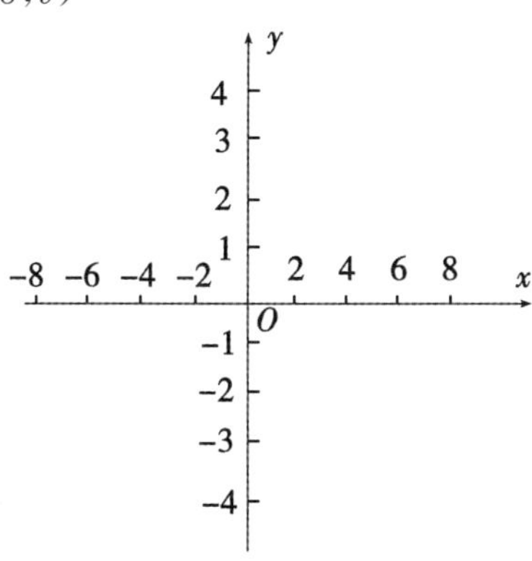

Level B

1. Read the following coordinates. Without graphing them, identify if they are linear relations.

a) $A(-1,-2)$, $B(0,0)$, $C(-1,2)$, $D(2,4)$

b) $A(-2,8)$, $B(-1,6)$, $C(0,4)$, $D(1,2)$

c) $A(6,-6)$, $B(5,-3)$, $C(4,0)$, $D(3,3)$

d) $A(-8,6)$, $B(-6,5)$, $C(-4,4)$, $D(-2,3)$

e) $A(-2,-4)$, $B(-1,-2)$, $C(0,1)$, $D(1,2)$

f) $A(-1,-2)$, $B(0,0)$, $C(1,2)$, $D(2,3)$

g) $A(3,9)$, $B(2,4)$, $C(1,1)$, $D(0,0)$

h) $A(-3,-9)$, $B(-2,-4)$, $C(-1,-1)$, $D(0,0)$

i) $A(1,-4)$, $B(2,-2)$, $C(2,0)$, $D(3,2)$

j) $A(-2,-4)$, $B(-1,-2)$, $C(1,4)$, $D(2,2)$

2. Find the point that is not on a straight line with others.

a) $A(0,1)$, $B(1,3)$, $C(2,5)$, $D(3,6)$

b) $A(-2,8)$, $B(-1,6)$, $C(0,4)$, $D(1,-2)$

c) $A(6,-6)$, $B(5,-3)$, $C(4,1)$, $D(3,3)$

d) $A(-8,6)$, $B(-5,5)$, $C(-4,4)$, $D(-3,3)$

e) $A(-2,-4)$, $B(-1,-2)$, $C(0,1)$, $D(1,2)$

f) $A(-2,5)$, $B(-1,7)$, $C(0,9)$, $D(1,10)$

g) $A(6,-3)$, $B(5,-4)$, $C(4,1)$, $D(3,3)$

h) $A(-1,-5)$, $B(0,-1)$, $C(1,3)$, $D(2,6)$

i) $A(1,-4)$, $B(2,-2)$, $C(4,2)$, $D(6,4)$

j) $A(-2,-4)$, $B(-1,-5)$, $C(1,-7)$, $D(2,-9)$

3. Find the missing value from a set of points on a straight line.

a) $A(-3,-7)$, $B(-2,-4)$, $C(-1,-1)$, $D(0, \underline{?})$

b) $A(1,3)$, $B(\underline{?},5)$, $C(5,7)$, $D(7, 9)$

c) $A(\underline{?},3)$, $B(3,5)$, $C(4,6)$, $D(5,7)$

d) $A(1,\underline{?})$, $B(4,6)$, $C(6,8)$, $D(7,9)$

e) $A(-3,-5)$, $B(-1,-1)$, $C(1,\underline{?})$, $D(3,7)$

f) $A(0,\underline{?})$, $B(2,4)$, $C(4,6)$, $D(6, 8)$

g) $A(2,4)$, $B(3,5)$, $C(\underline{?},8)$, $D(7,9)$

h) $A(\underline{?},4)$, $B(-1,5)$, $C(1,7)$, $D(2,8)$

4. Draw a graph and find the slope based on the following pairs of coordinates.

a) $A(-3,-3)$, $B(-2,-2)$, $C(-1,-1)$, $D(0,0)$

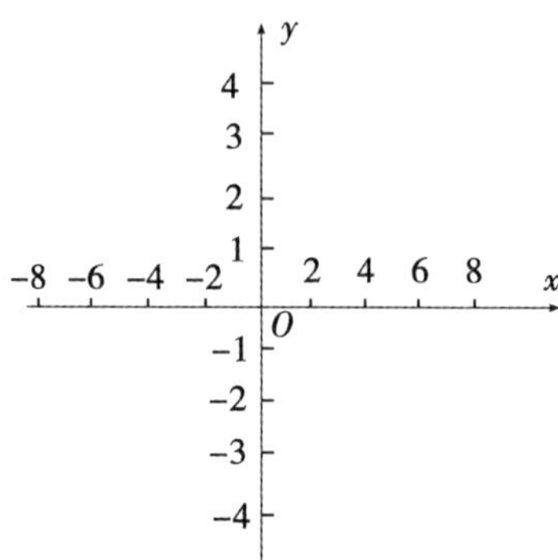

b) $A(0,2)$, $B(1,3)$, $C(2,4)$, $D(3,5)$

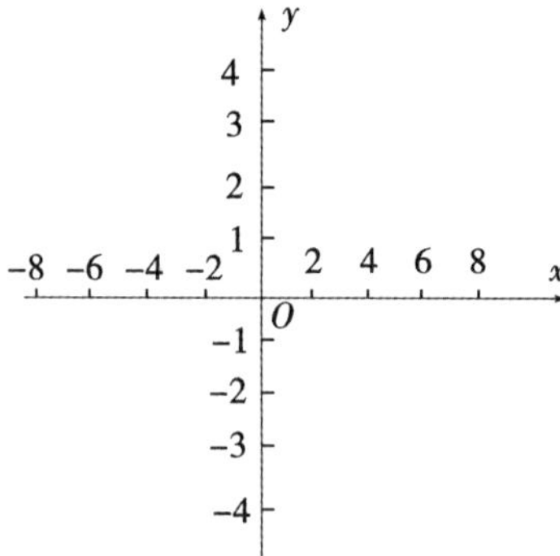

c) $A(-3,1)$, $B(-1,2)$, $C(1,3)$, $D(3,4)$

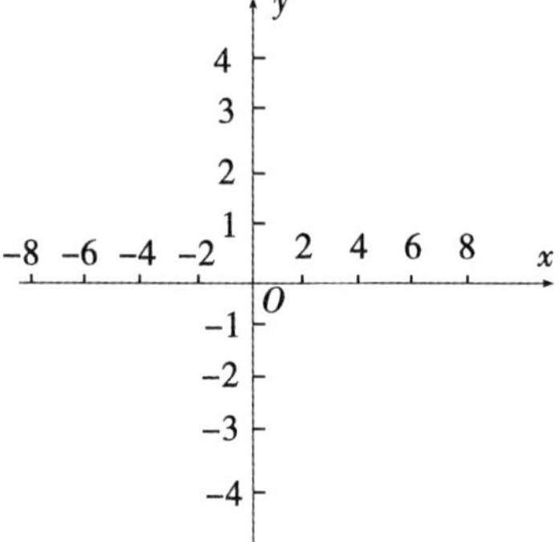

d) $A(-3,4)$, $B(-1,2)$, $C(1,0)$, $D(3,-2)$

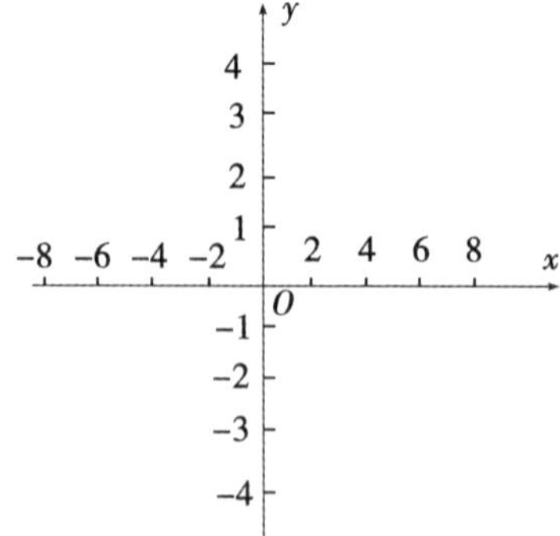

Level C

1. Tony is driving from Vancouver to Kamloops. The following are readings taken from the dashboard:

Fuel Remaining (gal)	10	7	5
Distance Driven (mile)	140	245	315

a) Determine the rate of fuel consumption.

b) How much gas was in the tank when Tony started driving?

c) Write the equation in the form $y=mx+b$. Indicate the meaning of each variable used.

2. It is "bike to work" week in the city where Aaron lives, and he decided to give biking a try. The following table shows the relation between time and distance travelled.

Time (minutes)	3	10	21
Distance (m)	900	3000	6300

a) How fast is Aaron travelling?

b) If his workplace is 9.3 km away from where he lives, how long does it take Aaron to bike to work?

c) Write the equation in the form $y=mx+b$. Indicate the meaning of each variable used.

Chapter 5 TEST

1. Find the value of the following expressions when $a=3$, $b=-3$ and $c=2$.

a) $\frac{2ab\times4}{5c-2b}$

b) $\frac{(2ac+2bc)\times5}{3a\times2c}$

c) $\frac{(3a-2b)\times3c}{ab\times4ac}$

d) $\frac{2a^2(b^2+c^2)}{6abc}$

e) $\frac{c^2-2a^2}{a^2+bc^2}$

f) $\frac{(b^2+c-2a)a}{a^2(b^2+c^2)}$

2. Based on the coordinates, draw a graph and identify if it is a linear relation.

a) $A(-2,-4)$, $B(-1,-2)$, $C(0,0)$, $D(1, 2)$

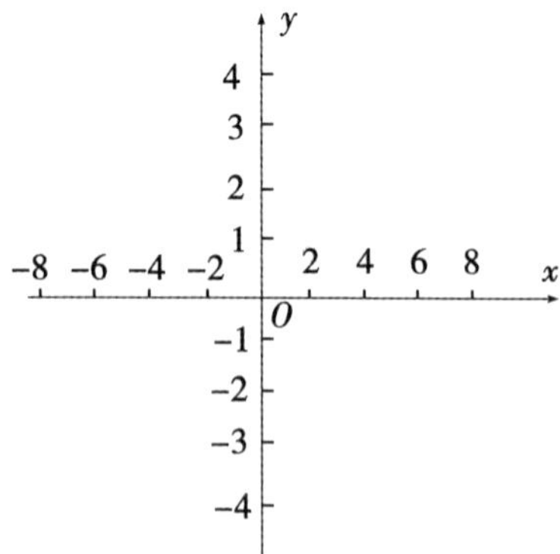

b) $A(1,5)$, $B(2,2)$, $C(3,0)$, $D(-2,-3)$

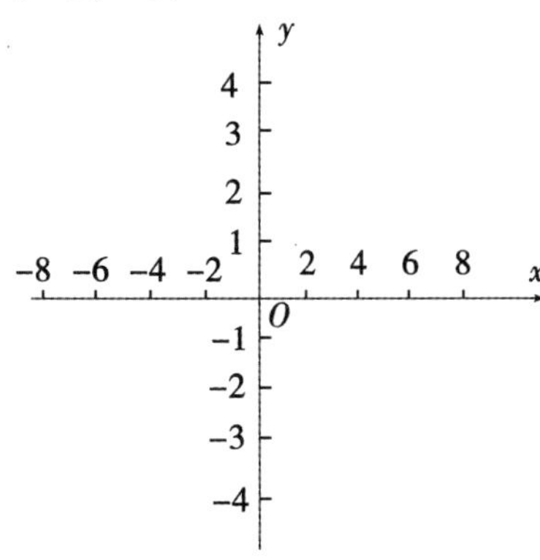

3. Draw graphs of the following equations.

a) $y=2x+1$ **b)** $y=-x-2$ **c)** $x=2$ **d)** $y=x^2$

a)

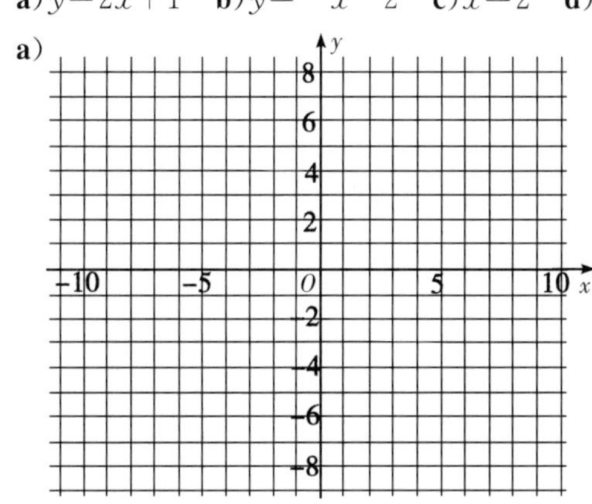

b)

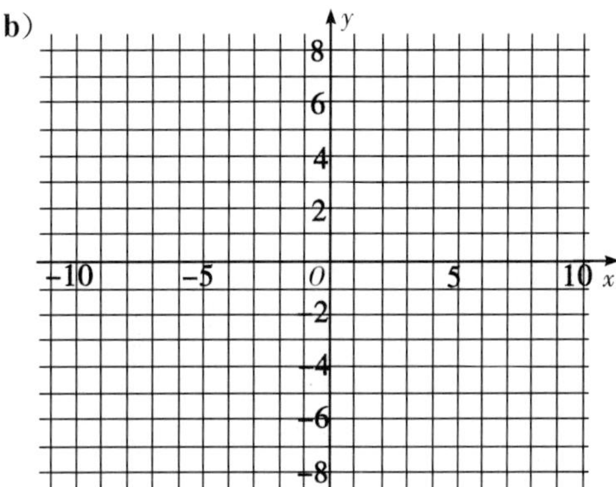

c)

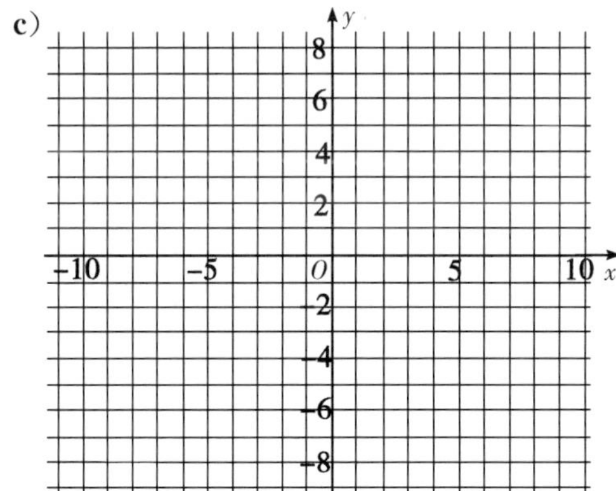

d)

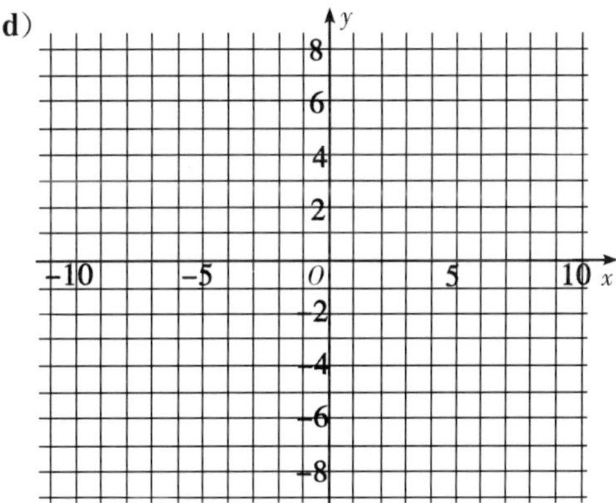

4. Complete the following tables. Find the formula that represents the relation between x and y, and draw graphs.

a)

x	-3	-2	0	1	2	3	4
y	2	0	-4	-6			

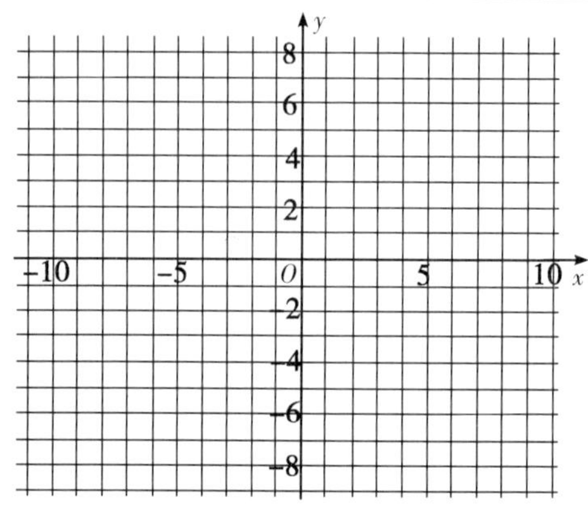

b)

x	-8	-5	-2	1	4	7	10
y	3	2	1	0			

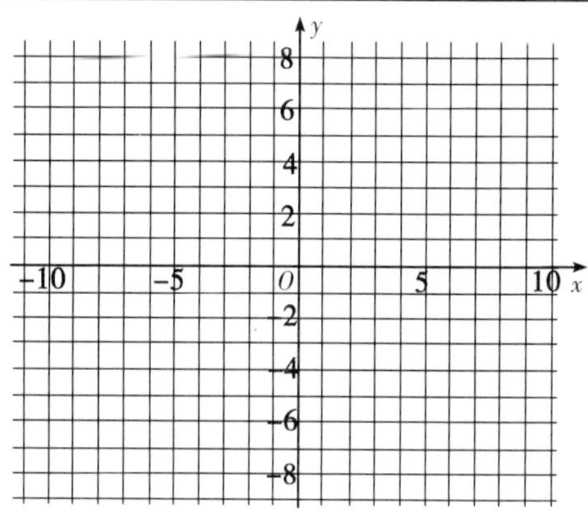

5. Draw graphs and find the slopes based on the following pairs of coordinates.

a) $A(-5,-5)$, $B(-3,-3)$, $C(-1,-1)$, $D(0, 0)$

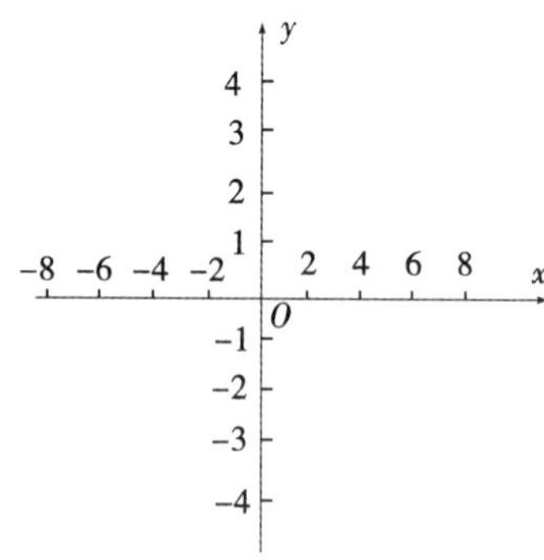

b) $A(-3,5)$, $B(-2,3)$, $C(-1,1)$, $D(0,-1)$

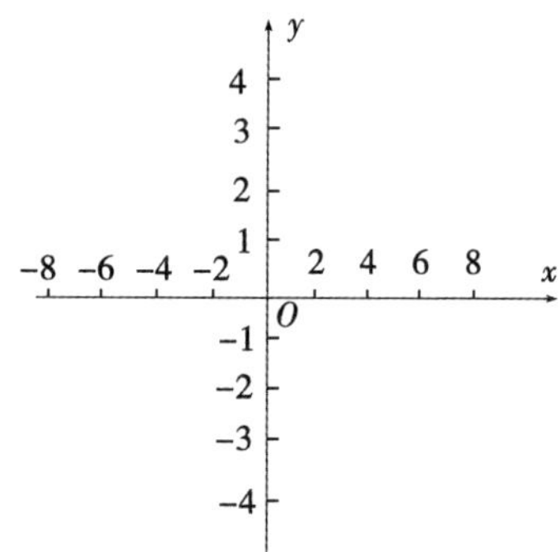

c) $A(-4,1)$, $B(-2,2)$, $C(-1,3)$, $D(2,4)$

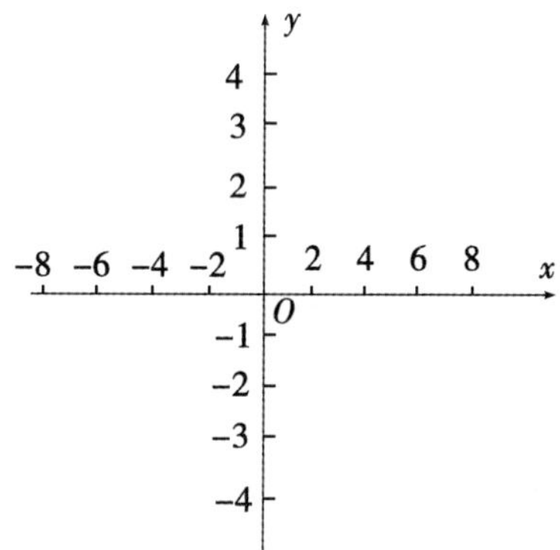

d) $A(-2,4)$, $B(0,2)$, $C(2,0)$, $D(4,-2)$

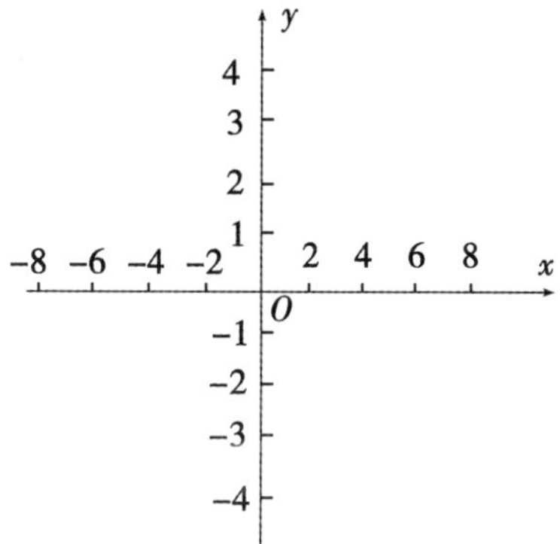

6. There is a fixed cost of \$50 to design a personal business card. It then costs \$0.21 to print one business card.

a) Determine the total cost of making 300 business cards.

b) How many business cards can be printed for a total cost of \$500?

CHAPTER 6

Linear Equations

6. 1 Combining Like Terms and Removing Brackets

6. 2 Solving Simple Linear Equations

6. 3 Solving Complicated Equations

Chapter 6 TEST

Chapter 6 Linear Equations

6.1 Combining Like Terms and Removing Brackets

To solve a linear equation, the basic method is to change a linear equation to the $Ax=C$ form and find the value of x. However, there are other definitions that need to be covered first.

Polynomial

A **polynomial** is a set of terms combined by "+" and/or "−" signs. For example, $2x^4+6x^3-5x^2+7x-9$.

Term

A **term** is a basic unit in a polynomial, which is separated from other terms by "+" and "−". For example, the polynomial $2x^4+6x^3-5x^2+7x-9$ has 5 terms, which are $2x^4$, $6x^3$, $-5x^2$, $7x$ and -9.

Degree of a Term

The **degree of a term** is the sum of the exponents of all variables in the term. For example, in $5a^2b^3c$, the degree of the term is $2+3+1=6$ (note that c is equivalent to c^1).

Coefficient

A **coefficient** is the number by which a term's variables are multiplied. For example, in $5x^4$, 5 is the coefficient.

Constant Term

A **constant term** is a term without a variable. For example, in the polynomial $2x^4+6x^3-5x^2+7x-9$, -9 is a constant term.

Like Terms

Like terms are terms with the same variables to the same exponents. For example, of $5a^2b^3$, $5a^2b^2$ and a^2b^3, only $5a^2b^3$ and a^2b^3 are like terms.

Combining Like Terms

Combining like terms is a very important step in solving an equation. When combining like terms, only the coefficients of the terms can be combined. The variables and exponents remain unchanged.

Example 1

Simplify.

$5a^2b^3+3a^2b^2-a^2b^3-a^3b^3$

▶ *Solution*:

$5a^2b^3+3a^2b^2-a^2b^3-a^3b^3=4a^2b^3+3a^2b^2-a^3b^3$

$5a^2b^3$ and $-a^2b^3$ are the only two like terms that can be combined.

Removing Brackets

To simplify an expression with a term in front of a bracket, multiply the term in the front to each of the terms inside the bracket, and then collect like terms.

Example 2

Simplify.

$-2x(3xy-2y+3x)+10x^2$

▶ *Solution*:

Expand the expression and combine like terms.

Multiply each of the terms inside of the brackets by the "$-2x$" outside of the brackets.

$-2x(3xy-2y+3x)+10x^2=-6x^2y+4xy-6x^2+10x^2$

$=-6x^2y+4xy+4x^2$ (Combine the like terms: $-6x^2$, $+10x^2$)

Exponent Rules

$a^m\times a^n=a^{m+n}$, $\dfrac{a^m}{a^n}=a^{m-n}$

Example 3

Simplify the following polynomials.

a) $(12xy^2)(-4x^4y^3)$

b) $\dfrac{-6a^3b^4a^2}{-24a^2b^3}$

▶ *Solution*:

a) $(12xy^2)(-4x^4y^3)=12\cdot(-4)x^{1+4}y^{2+3}=-48x^5y^5$

b) $\frac{-6a^3b^4a^2}{-24a^2b^3}=\frac{-6}{-24}a^{3+2-2}b^{4-3}=\frac{1}{4}a^3b$

Even if the exponent is a fraction or decimal, the rules of exponents still apply.

Example 4

Simplify.

$(x^{\frac{1}{2}}y^{\frac{1}{3}}z^{0.2})(x^{\frac{1}{3}}y^{\frac{1}{5}}z^{0.1})$

▶ *Solution*:

By applying the rules of exponents, we can obtain the following:

$(x^{\frac{1}{2}}y^{\frac{1}{3}}z^{0.2})(x^{\frac{1}{3}}y^{\frac{1}{5}}z^{0.1})$

$=(x^{\frac{1}{2}+\frac{1}{3}})(y^{\frac{1}{3}+\frac{1}{5}})(z^{0.2+0.1})$

$=(x^{\frac{3+2}{6}})(y^{\frac{5+3}{15}})(z^{0.3})$

$=x^{\frac{5}{6}}y^{\frac{8}{15}}z^{0.3}$

Practice

■Level A

Fill in the following table.

Polynomial	Number of Terms	Highest Degree of Terms	Coefficients	Like Terms	Constant Term
$2xy-5x+3xy-4$	4	2	2 and 3 for xy, -5 for x	$2xy$ and $3xy$	-4
$3x+5xy+4xy-2$					
$2x^2-x^2+3x-x$					
$2a^2b-2c^2+3a^2-2b$					
$3x^2y^2-6y^2+3y^2-uvxyz$					
$2.1\times10^8+2.5a\times10^7$					

Level B

1. Simplify the following polynomials. (Combine like terms)

a) $2xy-5x+3xy-4$

b) $3x+3y+2x+2y$

c) $3x+5xy+4xy-2$

d) $4x-4y+3x-3y$

e) $3x^2-3x+x^2-x$

f) $2x^2-3x+10^2-5x+10$

g) $3x^2y-3xy+3x^2y-3xy$

h) $4x^2y+4xy^2+4x^2y-4xy^2$

2. Simplify the following expressions. (Remove brackets and combine like terms)

a) $5y-2(5+3xy-4x)$

b) $3x(x+2y+3x)-2(x^2-xy)$

c) $3x-3x(15xy+4y)-12xy$

d) $y(x+2y+2x)-x(y-2x-2y)$

e) $5x^2-3x^2y-(7x^2-3y+2xy)$

f) $a^2b-5b^2a-b(3ab+3a^2-2a^2b)$

g) $-2x(3xy-3y)-2y(x-4x-4)$

h) $-a(ab-5b^2a)-b(5ab+a^2+5a^2b)$

i) $2.1\times10^8+1.2\times10^7+2a\times10^7+3a\times10^6$

j) $2b\times10^2+2a\times10^2+2a\times10^3+3b\times10^3+2\times10^2$

3. Simplify the following expressions. (Remove brackets and combine like terms)

a) $(12x^2y^4)(-4x^4y^3)$

b) $(-2^2x^3y^4)(-3x^4y^3)$

c) $(-3^2x^2y^2)^2(2^2x^3y^3)^2$

d) $\dfrac{36x^2y^4x^2}{-24y^2z^4}$

e) $\dfrac{-6a^3b^4a^2}{-24a^2b^3}$

f) $\dfrac{3x^2(2y^4-4y^2z^4)}{-24y^2z^4}$

g) $\dfrac{(-2b\times10^2+2a\times10^2)(2a\times10^3+3b\times10^3)}{(3b\times10^3+3a\times10^3)(3a\times10^2+3b\times10^2)}$

■ Level C

Simplify.

a) $4x^2y^2(x^{\frac{2}{3}}y^{\frac{2}{5}})+\frac{4}{9}(xy)y^{\frac{7}{5}}x^{\frac{5}{3}}$

b) $(0.4x^{\frac{2}{3}}y^{\frac{5}{2}}z^{\frac{3}{4}})(\frac{4}{5}x^{\frac{4}{5}}y^{\frac{4}{3}}z^{\frac{2}{3}})+(0.7x^{0.5}y^{\frac{1}{5}}z^{0.25})(\frac{1}{10}x^{\frac{1}{3}}y^{0.75}z^{\frac{1}{3}})-(\frac{2}{3}x^{0.4}y^{\frac{1}{6}}z^{\frac{5}{4}})(\frac{3}{4}x^{\frac{16}{15}}y^{\frac{11}{3}}z^{\frac{1}{6}})$

c) $\dfrac{7x^{\frac{5}{3}}y^{\frac{2}{3}}z^{\frac{7}{5}}}{19x^{\frac{1}{7}}y^{\frac{1}{5}}z^{0.4}}\times\dfrac{91x^{\frac{2}{9}}y^{\frac{3}{5}}z^{0.5}}{13x^{\frac{5}{3}}y^{0.8}z^{\frac{1}{4}}}$

d) $\dfrac{13x^{\frac{3}{2}}y^{\frac{1}{3}}z^{0.1}}{\dfrac{51x^{0.3}y^{0.6}z^{0.8}}{3x^{\frac{1}{5}}y^{\frac{1}{2}}z^{\frac{2}{3}}}}$

6.2 Solving Simple Linear Equations

Definition of Linear Equation

If the exponents of the variables in an equation are all 1, the equation is a **Linear Equation.** For example: $3x+15=27$, $5x-2y=19$.

Addition(Subtraction) Property of Equality

Adding or subtracting the same quantity to both sides of an equation makes an equivalent equation.

For example, if we have an equation, $x=18$

then

$x+3=18+3$

or

$x-5=18-5$

Term Moving Property of Equality

When a term on one side of the equals sign of a balanced equation moves to the other side of the equals sign, the opposite sign should be put in front of this term and this equation is still balanced. This property is actually a variation of the **Addition (Subtraction) Property.**

For example, if we have an equation:

$x+a=18$

then

$x=18-a$

Multiplication(Division) Property of Equality

Multiplying or dividing both sides of an equation by the same non-zero quantity makes an equivalent equation.

For example, if we have an equation:

$x=5$

then

$3\times x=3\times 5$

or

$\frac{x}{4}=\frac{5}{4}$

These properties are used to solve linear equations. The Multiplication(Division)

Property is used to solve equations of the form $Ax=C$ equations. The Addition (Subtraction) Property and the Term Moving Property are used to solve equations of the form the $Ax+B=C$.

In the next few examples, we will illustrate how to use different methods to solve linear equations including addition, subtraction, multiplication and division. On some occasions, more than one method can be used.

Example 1

Solve the equation $x+8=15$.

▶ *Solution* 1:

We can solve this linear equation by subtracting 8 on both sides. Then we obtain:

$x+8-8=15-8$

$x=7$

▶ *Solution* 2:

We can also solve this linear equation by using the Term Moving Property. We obtain the same answer:

$x+8=15$

$x=15-8$

$x=7$

Example 2

Solve the equation $x-12=18$.

For equations involving subtraction, we also have two methods that can be used to solve.

▶ *Solution* 1:

By adding 12 on both sides, we obtain:

$x-12+12=18+12$

$x=30$

▶ *Solution* 2:

By using the Term Moving Property, we obtain:

$x-12=18$

$x=18+12$

$x=30$

Example 3

Solve the equation $3x=5$.

▶ *Solution*:

We can apply the Division Property by dividing both sides by 3.

$$\frac{3x}{3}=\frac{5}{3}$$

$$x=\frac{5}{3}$$

Example 4

Solve the equation $\frac{x}{4}=5$.

▶ *Solution* 1:

We can apply the Multiplication Property by multiplying both sides by 4.

$$\frac{x}{4}\times4=5\times4$$

$$x=20$$

▶ *Solution* 2:

The other way to solve this question is the following:

$$\frac{x}{4}=5$$

$$\frac{1}{4}\times x=5$$

We then divide both sides by one quarter.

$$\frac{\frac{1}{4}x}{\frac{1}{4}}=\frac{5}{\frac{1}{4}}$$

$$x=\frac{5}{\frac{1}{4}}$$

$$x=20$$

Example 5

Solve the equation $\frac{2x}{3}=\frac{3}{5}$.

▶ *Solution*:

As with the previous examples, there are two methods to solve this problem. Only one will be shown here. You can try the other method.

$\frac{3}{2}\times\frac{2}{3}x=\frac{3}{2}\times\frac{3}{5}$

$x=\frac{9}{10}$

Example 6

Solve the equation $5(x-2)=3(x+4)$.

▶ *Solution*:

Expand the equation and combine like terms.

$5x-10=3x+12$

$5x-3x=10+10$

$2x=22$

$x=11$

Practice

■Level A

1. Solve the following equations.

a) $2.3x=6.9$ **b)** $0.5x=1.5$ **c)** $1.2y=1.44$ **d)** $2.1y=6.3$

e) $0.2x=5$ **f)** $0.15x=1.5$ **g)** $0.8y=4$ **h)** $4y=1.2$

i) $\frac{2}{3}x=18$ **j)** $\frac{2}{5}x=4$ **k)** $\frac{5}{6}y=\frac{1}{3}$ **l)** $\frac{y}{5}=\frac{2}{9}$

m) $\frac{5}{6}x=-30$ **n)** $-\frac{2}{3}x=10$ **o)** $-\frac{3}{4}y=-\frac{1}{2}$ **p)** $-\frac{3y}{5}=\frac{1}{25}$

2. Tom plans to read a novel of 620 pages during July. How many pages will he read each day? Set up an equation and solve it.

Level B

1. Solve the following equations.

a) $5x-5.5=2.5$

b) $0.9x+1.5=2.4$

c) $y-(20-5)=-17$

d) $5y-(3y+4)=16$

e) $\frac{5}{24}-(-5x)=10$

f) $\frac{3}{4}y-2=12$

g) $-15=4m+5$

h) $10-6v=-104$

i) $-9x-13=-103$

j) $-10=-10(k-9)$

k) $-243=-9(10+x)$

l) $3(x-4)=48$

m) $6x-2(x-3)=18$

n) $3(x-2)-4(x-3)=100$

o) $10x-2(2x-2^2)=x+10$

2. Set up a variable (x) and an equation to solve the following questions.

a) Five more than one half of a number is 25. Find the number.

b) 40 equals to the sum of 10 and a quarter of a number. Find the number.

c) 200 equals to 75 less the sum of 25 and a number. Find the number.

d) Three times a number is 5 less than 50. Find the number.

e) 120 increased by the quotient of a number divided by 9 equals to 125. Find the number.

3. 32 mL of Brand M cinnamon was made by combining 24 mL of Indonesian cinnamon, which costs \$1.9/mL, with 8 mL of Thai cinnamon, which costs \$1.1/mL. Find the cost per mL of the mixture.

■Level C

Isolate the indicated variable.

a) $A=\frac{(a+b)h}{2}$, find a.

b) $d=V_i t+\frac{1}{2}at^2$, find a.

c) $\frac{1}{R_t}=\frac{1}{R_1}+\frac{1}{R_2}+\frac{1}{R_3}$, find R_2.

d) $d_0=\frac{fd_i}{d_i-f}$, find d_i.

e) $T=2\pi\sqrt{\frac{m}{k}}$, find m.

6.3 Solving Complicated Equations

Basic Steps to Solving a Linear Equation

1. Remove brackets and denominators.

2. Combine like terms.

3. By using the Addition Subtraction Property or the Term Moving Property, move all the terms with a variable to one side of the equal sign, and move all the terms without a variable to the other side of the equal sign.

4. After simplifying, apply the Multiplication Division Property to solve the equation.

5. Check the answer by substituting it into the original equation. If both sides of the equation are equal, the answer is correct.

Example 1

Solve the equation: $2x+3=5$.

▶ *Solution*:

$2x+3=5$

$2x=5-3$

(Use the Term Moving Property to move 3 to the right side and change the sign to "$-$"; then use the Division Property on both sides of the equal sign by dividing both by 2.)

So $x=1$.

Check: $2\times(1)+3=5$

(Substitute the answer $x=1$ into the original equation $2x+3=5$. The amount on the left side of the equal sign is equal to the amount on the right side, so the solution is correct.)

Example 2

Solve the equation $\frac{2x}{5}+1=\frac{2x}{3}-\frac{1}{5}$

▶ *Solution*:

Multiply the Least Common Denominator by all terms so that we can have an equation without a denominator.

$15(\frac{2}{5}x+1)=15(\frac{2x}{3}-\frac{1}{5})$

$6x+15=10x-3$

$6x-10x=-3-15$(move all the terms with variables to one side, and move all the constants to the other side)

$-4x=-18$

$x=\frac{-18}{-4}=\frac{9}{2}$

Check:

$\frac{2}{5}\times\frac{9}{2}+1=\frac{2}{3}\times\frac{9}{2}-\frac{1}{5}$

$\frac{9}{5}+1=\frac{9}{3}-\frac{1}{5}$

$\frac{14}{5}=3-\frac{1}{5}=\frac{15}{5}-\frac{1}{5}=\frac{14}{5}$(true)

Example 3

Solve the equation $4(2-\frac{3}{5}x)=24-2(\frac{4}{5}x-2\frac{2}{5})$.

▶ *Solution*:

$4(2-\frac{3}{5}x)=24-2(\frac{4}{5}x-2\frac{2}{5})$

$4(10-3x)=120-2(4x-12)$

(Multiply the Least Common Multiple on both sides of the equals sign so that we can have an equation without a denominator.)

$40-12x=120-8x+24$(Remove brackets.)

$40-120-24=-8x+12x$

(By using the Term Moving Property, move all terms with the variable x to the right side of the equals sign, and move all the constants to the left side of the equals sign.)

$-104=4x$ (Combine the like terms, and solve C=Ax.)

$x=-26$ (Using the Division Property, divide both sides of the equals sign by 4.)

Check: $4\times(2-\frac{3}{5}\times(-26))=24-2\times(\frac{4}{5}\times(-26)-2\frac{2}{5})$

Basic Steps to Solving Word Problems

1. Define a variable, such as x, as the unknown value in the given problem.

2. Set up an equation.

3. Solve the equation and write out the answer in a statement such as "The number is..."

4. Substitute the solution into the original equation to check.

Example 4

200 equals to 100 less two times the sum of a number and 25. What is the number?

▶ *Solution*:

Let x be the unknown number.

$100-2(x+25)=200$

$100-2x-50=200$

$50-200=2x$

$-150=2x$

$x=-75$

The number is -75.

Check:

$100-2\times[(-75)+25]=200$

$100-2\times(-50)=200$

$100-(-100)=200$

$200=200$

The solution is correct.

Example 5

Nick is 45 years old. Nick's age is 1.5 times Tony's age, and Tony's age is $\frac{5}{4}$ of Steven's age. How old are Steven and Tony?

▶ *Solution*:

Let Steven's age be x.

Tony's age: $x\times\frac{5}{4}$

Nick's age: $(x\times\frac{5}{4})\times1.5=45$

$(x\times\frac{5}{4})=\frac{45}{1.5}$ (Use the Multiplication Property)

$x=\frac{45}{1.5}\times\frac{4}{5}$ (Use the Multiplication Property again)

$x=\frac{9}{1.5}\times\frac{4}{1}=24$ (Simplify before multiplying & solve for x)

Steven is 24 years old and Tony is $24\times\frac{5}{4}=30$ years old.

Example 6

Solve the following linear equation.

$7(\frac{3}{4}x+6x-5)+9=2$

▶ *Solution*:

$7(\frac{3}{4}x+6x-5)=2-9=-7$ (Term Moving Property)

$(\frac{3}{4}x+6x-5)=\frac{-7}{7}=-1$ (Multiplication Property)

$\frac{3}{4}x+6x=-1+5=4$ (Get rid of brackets & Term Moving Property)

$\frac{3}{4}x+\frac{24}{4}x=\frac{27}{4}x=4$ (Combine like terms)

$x=4\times\frac{4}{27}=\frac{16}{27}$ (Multiplication Property x)

Example 7

Moe, Larry and Harry shared a pizza for lunch. Larry ate half as much as Moe did, but twice as much as Harry did. When they were done, there was still one slice left. Assume each slice is one eighth of the pizza. How much did each person eat?

▶ *Solution*:

If x represents the amount of pizza Larry ate, then Moe ate $2x$ and Harry ate $\frac{1}{2}x$.

$2x+x+\frac{1}{2}x+\frac{1}{8}=1$ (Let the entire pizza be 1)

$2x+x+\frac{1}{2}x=1-\frac{1}{8}=\frac{7}{8}$ (Term Moving Property)

$\frac{4}{2}x+\frac{2}{2}x+\frac{1}{2}x=\frac{7}{2}x=\frac{7}{8}$ (Combine like terms)

$x=\frac{7}{8}\times\frac{2}{7}=\frac{1}{4}$ (Multiplication Property)

Larry ate $\frac{1}{4}$ of a pizza, or $\frac{\frac{1}{4}}{\frac{1}{8}}=\frac{1}{4}\times\frac{8}{1}=2$ slices; Moe ate $\frac{1}{2}$ of a pizza, or 4 slices; Harry ate $\frac{1}{8}$ of a pizza, or 1 slice.

Example 8

Solve the following linear equation.

$3(x+5)-9x=1+7x$

▶ *Solution*:

$3x+15-9x=1+7x$ (Remove brackets)

$3x-9x-7x=1-15$ (Isolate the variable by moving all terms with x to one side)

$-13x=-14$ (Combine like terms)

$x=\frac{-14}{-13}=\frac{14}{13}$ (Division property)

Example 9

Daniel, Christopher and Philip are trying to pool their money together in order to rent a car. The amount of money Christopher forks out is \$50 less than double of what Daniel chips in. In addition, Philip pays half of \$25 less than the sum of Daniel's and Christopher's costs. If they pooled \$475 altogether, how much did each person pay?

▶ *Solution*:

Let the amount Daniel pays be x. Then Christopher pays $2x-50$, and Philip pays $\frac{x+(2x-50)-25}{2}$

$x+(2x-50)+\frac{x+(2x-50)-25}{2}=475$ (Set up the equation)

$2x+2(2x-50)+x+(2x-50)-25=950$ (Multiply every term by 2)

$2x+4x-100+x+2x-50-25=950$ (Remove brackets)

$2x+4x+x+2x=950+100+50+25$ (Term Moving Property)

$9x=1125$ (Simplify)

$x=125$ (Multiplication Property)

Daniel pays \$125, Christopher pays $\$125\times2-50=\$200$, and Philip pays $\frac{\$125+(\$125\times2-50)-25}{2}=\$150$.

Practice

■Level A

1. Solve the following equations.

a) $5x=75$　　b) $-23x=-115$　　c) $21x=-168$

d) $-14x=126$　　e) $\frac{x}{12}=24$　　f) $-\frac{3}{5}x=12$

g) $-\frac{1}{10}x=-\frac{1}{5}$　　h) $-\frac{5}{6}x=1\frac{1}{2}$　　i) $3\frac{x}{2}+2=24$

j) $1\frac{1}{2}x-2\frac{1}{3}=5$ k) $\frac{x}{2}+2x=24$ l) $\frac{x}{2}-12=4x$

m) $15-\frac{x}{3}=44$ n) $0.05x-3=1-0.5x$ o) $-0.25y-1=0.25y-1$

2. The perimeter of a rectangle is 48 cm. If the length of one side is 6 cm, what is the length of the other side?

■Level B

1. Solve the following equations.

a) $3x-15+2x=10x$ b) $2y+20-7y=15y-5$

c) $0.5y+1.5=2.5-2.5y$ d) $2.8x+2.6=-1.2x+12.6$

e) $3(2x-5)=2(5x+4)$ f) $2(3y+2y)-3(y-5)=3y+5$

g) $\frac{1}{2}m+\frac{3}{5}m-2=(-2)(3-\frac{1}{2}m)$ h) $-2(\frac{1}{3}n+\frac{1}{4}n-2)=2-\frac{1}{3}n$

i) $(-\frac{1}{2}x)(-1.2)-2(\frac{3}{4}x-2)=-4.8$ j) $(-\frac{1}{3}x)(1.8)-3(\frac{x}{4}-3)=4(-4.8+1.2x)$

2. Bob started travelling to US to go shopping. Mary left 1. 8 hours later and travelled 20 km/h faster than Bob in an effort to catch up to him. After 8. 55 hours Mary finally caught up to him. Find Bob's average speed.

3. Jack can paint a big room in ten hours. Wilson can paint the same room in eight hours. If they worked together how long would it take them?

4. A boat traveled 168 km downstream and then went back upstream to its starting point. The trip downstream took 6 hours. The trip back took 7 hours. What is the speed of the boat in still water? What is the speed of the current?

■Level C

Solve the following equations.

a) $0.5(\frac{x}{4}+\frac{2}{3})=0.2(\frac{x}{3}-\frac{3}{4})$

b) $\frac{x+2}{4}-\frac{x-1}{2}=\frac{2}{3}$

c) $4+\frac{2}{x}=7+\frac{3}{x}$

d) $\frac{3}{2x}=\frac{4}{3x}-\frac{1}{2}$

Chapter 6 TEST

1. Simplify the following expressions. (Remove brackets and combine like terms)

a) $2x^3y-2xy+2x^3y-2xy$

b) $3x^2y+3xy^2+3x^2y-3xy^2$

c) $4x^2-4x^2y-(-5x^2-2y+3xy)$

d) $a^2b-5b^2a-b(3ab+3a^2-2a^2b)$

e) $-3x(2xy-2y)-3y(x-2x-5)$

f) $-2a(2ab-3b^2a)-2b(3ab+2a^2+4a^2b)$

2. Solve the following equations.

a) $5x-7(x-3)=20$

b) $2(x-3)-4(x-5)=120$

c) $9x-2(3x-3^2)=x-12$

d) $-8x-15=-95$

e) $-20=10(x-8)$

f) $-280=-7(10+x)$

g) $6(x-4)=48$

h) $5x-3(x-3)=16$

i) $2(x-3)-3(x-4)=100$

j) $20-\frac{x}{4}=42$

k) $0.09x-3=2-0.6x$

l) $-0.25y+2=0.25y-2$

3. Solve the following equations.

a) $4x-32+3x=15x$

b) $3y+29-7y=15y-5$

c) $0.3y+2.5=3.5-2.7y$

d) $1.8x-4.6=-3.2x+10.4$

e) $2(3x-5)=3(4x+4)$

f) $3(2y+3y)-2(y-6)=5y+15$

g) $-3(\frac{1}{3}m+\frac{2}{4}m-3)=2(4-\frac{1}{2}m)$

h) $-5(\frac{1}{2}n+\frac{1}{3}n-2)=-4(3-\frac{1}{5}n)$

i) $(-\frac{1}{4}x)(-1.6)-3(\frac{5}{6}x-2)=-3.6+x$

j) $(-\frac{1}{2}x)(1.8)-2(\frac{x}{3}-4)=-3(-5.4+1.2x)$

4. Set up a variable x and an equation to solve the following questions.

a) 6 more than one half of a number is 34. Find the number.

b) 50 is 10 more than a quarter of a number. Find the number.

c) 60 equals to 175 less the sum of 75 and a number. Find the number.

d) 3 times a number is 15 less than 60. Find the number.

e) 90 minus the quotient of a number divided by 5 equals to 175. Find the number.

5. Jason started travelling to US to visit his relatives. Jean left 0.6 hours after Jason and travelled 18 km/h faster than Jason in an effort to catch up to him. After 3.2 hours Jean finally caught up to Jean's. Find Jean's average speed.

6. Tony can paint a big room in eight hours. John can paint the same room in ten hours. If they worked together how long would it take them to paint the room?

7. A ship traveled 224 km downstream and then went back upstream to its starting point. The trip downstream took 7 hours. The trip back took 8 hours. What is the speed of the ship in still water? What is the speed of the current?

8. Simplify.

a) $\dfrac{12x^{\frac{7}{3}}y^{\frac{3}{8}}z^{\frac{4}{14}}}{25x^{\frac{1}{3}}y^{\frac{1}{5}}z^{0.6}} \times \dfrac{75x^{\frac{3}{4}}y^{\frac{3}{5}}z^{1.5}}{36x^{\frac{7}{3}}y^{0.2}z^{0.6}}$

b) $\dfrac{\dfrac{24x^{\frac{7}{2}}y^{\frac{1}{5}}z^{0.2}}{144x^{0.6}y^{0.5}z^{0.9}}}{6x^{\frac{1}{2}}y^{\frac{1}{3}}z^{\frac{2}{3}}}$

CHAPTER 7

Pythagorean Theorem, Combined Shapes and Tessellations

7. 1 Pythagorean Theorem

7. 2 The Perimeter and Area of Combined Shapes

7. 3 Tessellations

Chapter 7 TEST

Pythagorean Theorem, Combined Shapes and Tessellations

7.1 Pythagorean Theorem

Right Triangle

A **Right Triangle** is a triangle that has a right angle(90°). The longest side of a right triangle is called the **Hypotenuse.** Both of the two sides that form the right angle are called the **legs.**

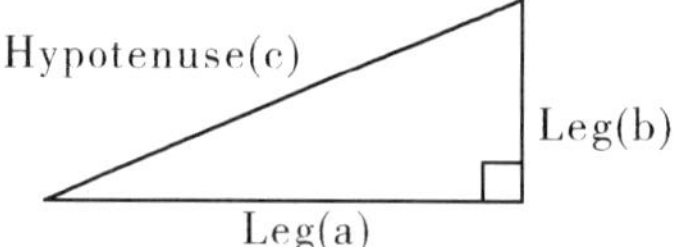

Definition of the Pythagorean Theorem

The **Pythagorean Theorem** states that the square of the hypotenuse of a right triangle equals to the sum of the squares of the other two sides. In the formula $c^2=a^2+b^2$, c is hypotenuse, a and b are the two legs of the right triangle.

Example 1

Find side x of the following right triangle.

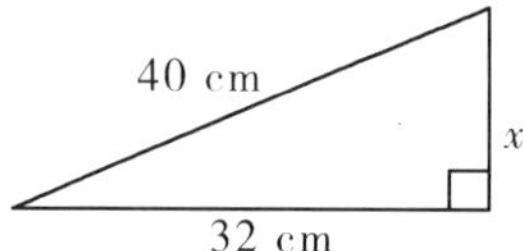

▶ *Solution*:

$c^2=a^2+b^2, a=\sqrt{c^2-b^2}=\sqrt{40^2-32^2}=24$ cm

Side x is 24 cm.

Example 2

Find the hypotenuse of the following right triangle.

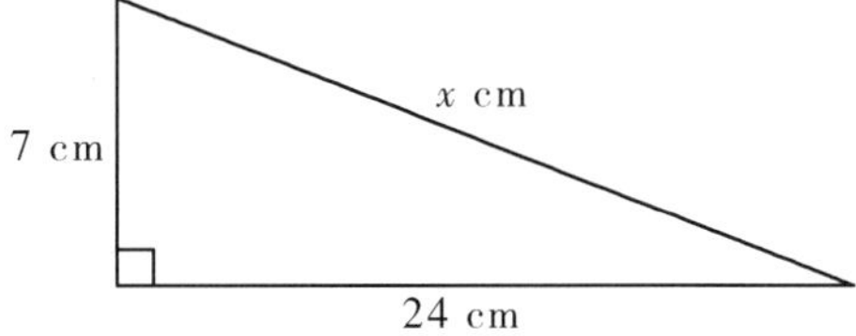

▶ *Solution*:

$x^2 = a^2 + b^2$

$x = \sqrt{a^2 + b^2}$

$x = \sqrt{7^2 + 24^2}$

$x = 25$

Example 3

Find the length of $\overline{CD}$ ($\overline{CD}$ is perpendicular to segment $\overline{AB}$). Round your answer to the nearest hundredth.

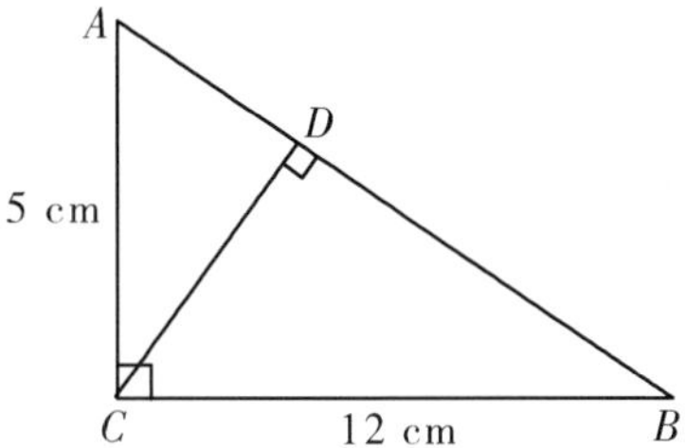

▶ *Solution*:

$\overline{AB}^2 = 5^2 + 12^2$

$\overline{AB} = \sqrt{5^2 + 12^2}$

$\overline{AB} = 13$

$\because$ Area $= \frac{5 \times 12}{2} = \frac{\overline{AB} \times \overline{CD}}{2}$

$\therefore$ $5 \times 12 = \overline{AB} \times \overline{CD}$

$\overline{CD} = \frac{5 \times 12}{\overline{AB}} = \frac{5 \times 12}{13} = \frac{60}{13} \approx 4.62$

Example 4

In the following diagram, $\overline{AO}=\overline{BO}=\overline{DO}=r=\sqrt{2}$, and $\overline{DO}$ is a perpendicular bisect of $\overline{AB}$. Find the length of $\overline{CD}$. Round your answer to the nearest hundredth.

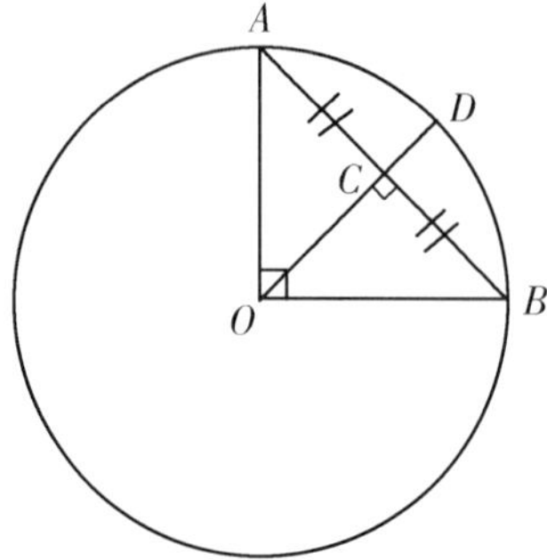

▶ *Solution*:

$\overline{AB}^2=\overline{AO}^2+\overline{BO}^2$

$\overline{AB}=\sqrt{\overline{AO}^2+\overline{BO}^2}$

$=\sqrt{(\sqrt{2})^2+(\sqrt{2})^2}$

$=\sqrt{2+2}$

$=\sqrt{4}$

$=2$

$\because$ Area of $\triangle ABO=\overline{AO}\times\overline{BO}\div 2=\overline{AB}\times\overline{CO}\div 2$

$\therefore$ $\overline{AO}\times\overline{BO}\div 2=\overline{AB}\times\overline{CO}\div 2$

$\overline{AO}\times\overline{BO}=\overline{AB}\times\overline{CO}$

$\overline{CO}=\dfrac{\overline{AO}\times\overline{BO}}{\overline{AB}}$

$=\dfrac{\sqrt{2}\times\sqrt{2}}{2}$

$=\dfrac{2}{2}$

$=1$

$\overline{CD}=\overline{DO}-\overline{CO}$

$=\sqrt{2}-1$

≈ 0.41

Common Pythagorean Triples for Right Triangles

While we can always use the Pythagorean Theorem to find the length of any side when given the other two sides, there are several common combinations of sides worth noting where all three sides are whole numbers. Knowing these combinations by heart will help speeding up the calculation.

Common Pythagorean Triples are in the format **leg: leg: hypotenuse.**

For example:

3 ∶ 4 ∶ 5

5 ∶ 12 ∶ 13

8 ∶ 15 ∶ 17

7 ∶ 24 ∶ 25

9 ∶ 40 ∶ 41

...

If we multiply each number of any triple by the same integer, the result will be another triple. For example, from the triple 3 ∶ 4 ∶ 5, we can have triples like 6 ∶ 8 ∶ 10, 30 ∶ 40 ∶ 50, etc, so the triple 3 ∶ 4 ∶ 5 is actually the **ratio** of the lengths, not necessarily the **actual** lengths.

Practice

■Level A

1. Find Side x in each of the following right triangles.

a)

x

10 cm

30 cm

b)

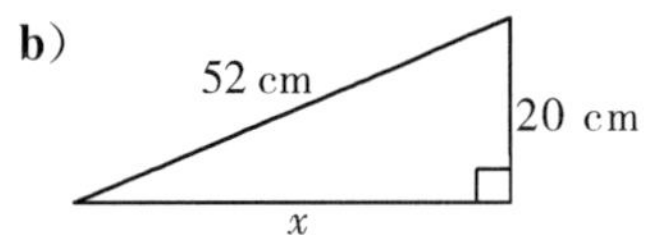

c)

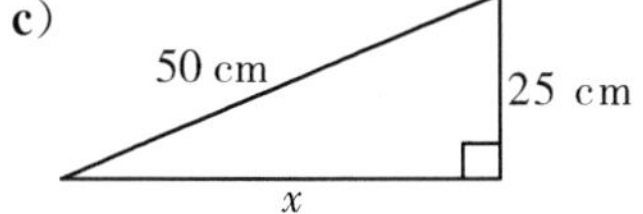

d)

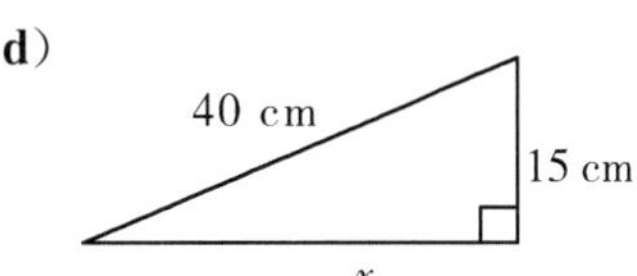

2. Find Side x in the following right triangles given $\angle A=90°$.

a)

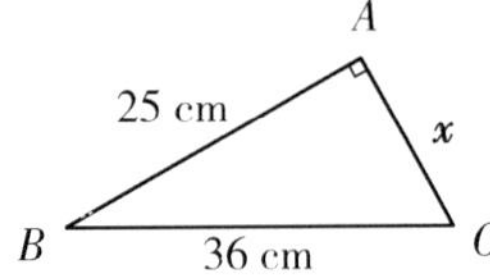

b)

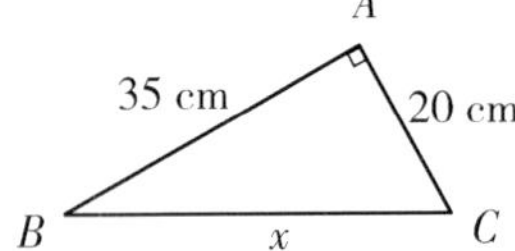

c)

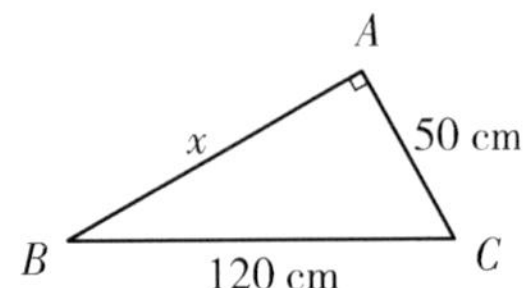

d)

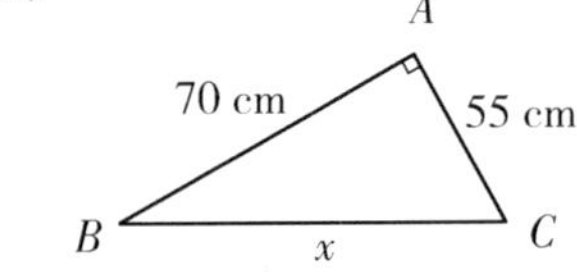

3. Find the height AD, perimeter and area of $\triangle ABC$.

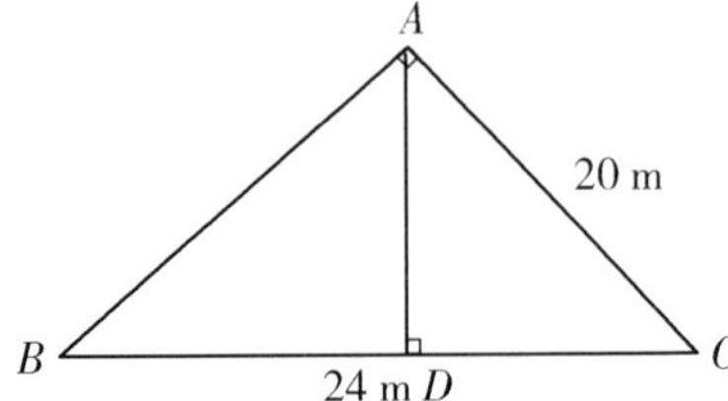

4. Find the length of Side AC for the following triangle, given $\angle A=90°$.

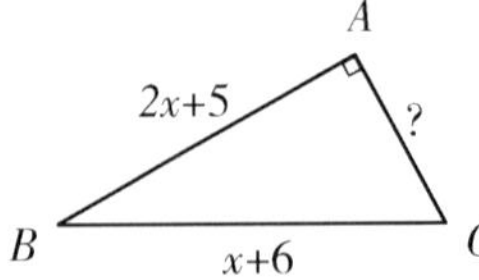

Level B

1. In the following figure, $AB=AC=13$, $BC=10$, D is the midpoint of BC, and DE is perpendicular to AB. How long is DE? Give your answer in exact form.

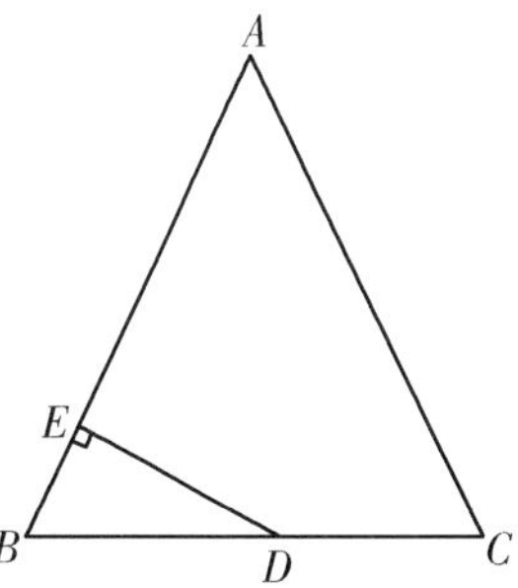

2. A rectangular paper $ABCD$ is folded over on BD, so Point C goes up to C'. If $AD=8$, and $AB=4$, how long is DE?

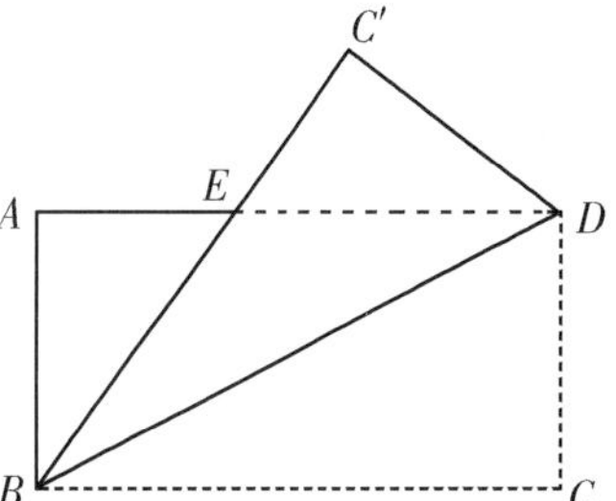

3. To get from Point A to Point B, you must avoid walking through a pond. To avoid the pond, you must walk 25 metres south and 38 metres east. To the nearest metre, how many metres would be saved if it were possible to walk through the pond?

4. A baseball diamond is a square with sides of 30 metres. What is the shortest distance, to the nearest tenth of a metre, between first base and third base?

5. A suitcase measures 0.6 m long and 0.45 m high. What is the diagonal length of the suitcase to the nearest tenth of a metre?

6. A computer monitor is listed as being 19 inches. This distance is the diagonal distance across the screen. If the screen measures 10 inches in height, what is the actual width of the screen to the nearest inch?

7. The old square floppy diskettes for computers measured 5 inches and $\frac{1}{4}$ inches on each side. What was the diagonal length of the diskettes to the nearest tenth of an inch?

8. Ms. Spires tells you that a right triangle has a hypotenuse of 26 cm and a leg of 10 cm. She asks you to find the other leg of the triangle. What is your answer?

9. Two joggers run 12 km north and then 6.5 km west. What is the shortest distance, to the nearest tenth of km, that they must travel to return to their starting point?

10. A truck drives 60 km east and then 46 km due north. How far is it from where it started? Round your answer to the nearest tenth of km.

11. When Tony drives to school, he picks up his friend Delavan. Delavan lives directly south of Tony. Their school is directly east of Tony's house. Tony usually drives an average of 50 km per hour. If Tony lives 20 km from Delavan, and Delavan lives 30 km from school, how much time would Tony save if he did not have to pick up his friend Delavan?

■Level C

1. A square has all four of its vertices on a circle. The area of the square is equal to 625 cm^2. What is the area of the circle? Give an exact value.

2. An isosceles right-triangle ABC is drawn in a circle. The diameter of the circle is $4\sqrt{2}$ cm. Determine the length of $\overline{DE}$. Give an exact value.

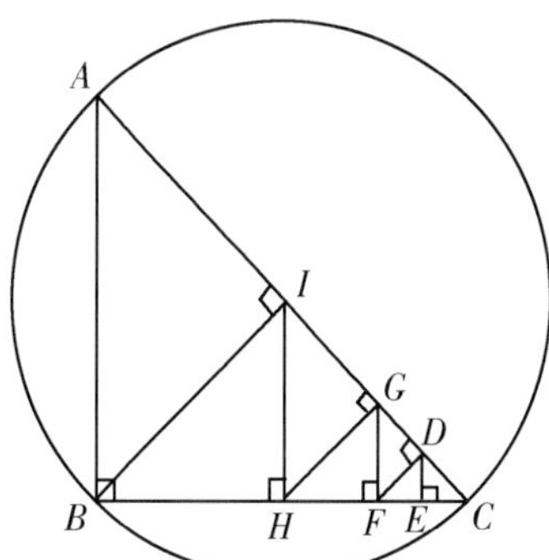

7.2 The Perimeter and Area of Combined Shapes

Shape Name	Graph	Perimeter	Area
Parallelogram	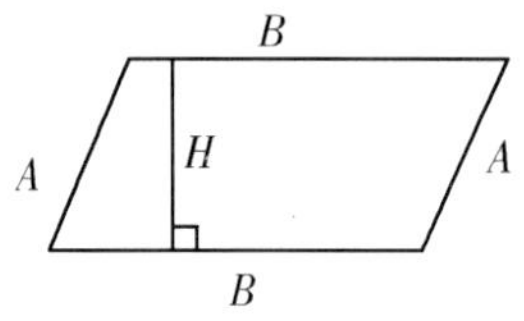	$P=2(A+B)$	$\text{Area}=HB$
Triangle	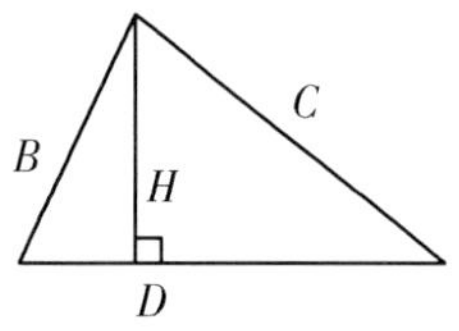	$P=B+C+D$	$\text{Area}=\frac{DH}{2}$
Trapezoid	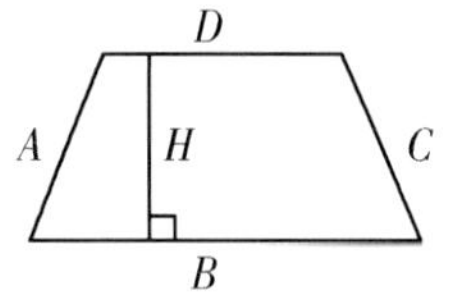	$P=A+B+C+D$	$\text{Area}=\frac{(B+D)H}{2}$
Isosceles Triangle	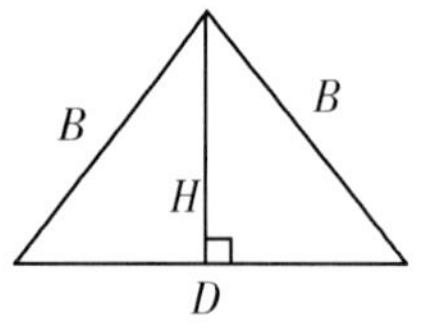	$P=2B+D$	$\text{Area}=\frac{DH}{2}$
Circle	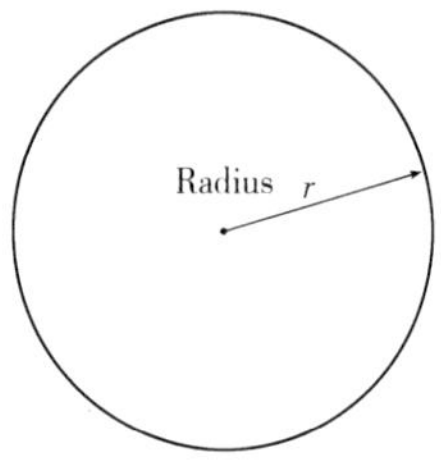 Radius r	$C=2\pi r$	$\text{Area}=\pi r^2$
Sector	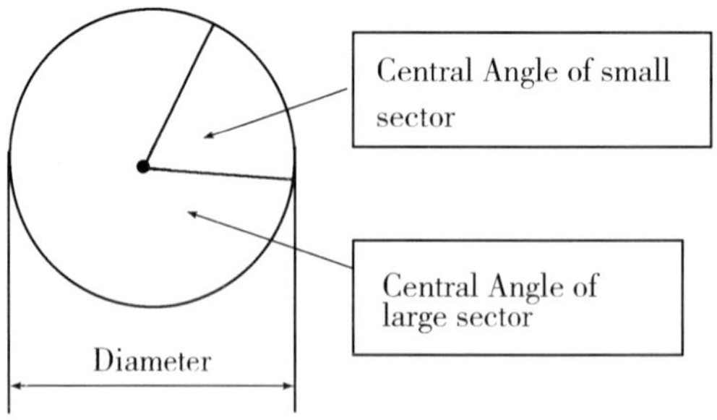 Central Angle of small sector; Central Angle of large sector; Diameter		

Sector Perimeter	$C=2\pi r\frac{Central\ Angle}{360^{\circ}}$ (The unit of the central angle is in degrees) $+2r$
Sector Area	$A=\pi r^{2}\frac{Central\ Angle}{360^{\circ}}$ (The unit of the central angle is in degrees)

Example

Find the perimeter and the area of the following figure. Round the answers to the nearest hundredth.

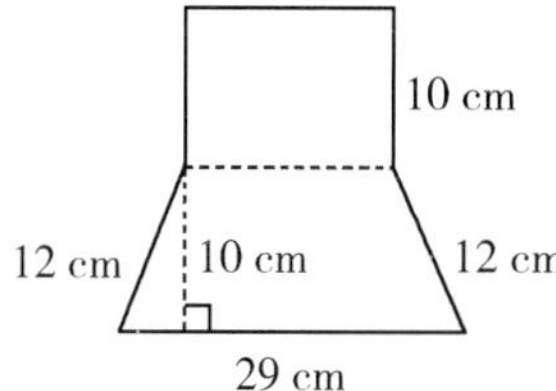

▶ *Solution*:

The shape is a combination of a rectangle and a trapezoid.

We need to calculate the length of the top side of the trapezoid (the line where the rectangle and the trapezoid meet):

$29-2\sqrt{12^{2}-10^{2}}=15.73$ cm

So, the perimeter is: $15.73+29+2\times10+2\times12=88.73$ cm.

The area is: $15.73\times10+\frac{10}{2}(29+15.73)=380.95$ cm^2

The perimeter is 88.73 cm. The area is 380.95 cm^2.

Practice

■Level A

1. Find the perimeter and the area of the following figure and round the answers to the nearest hundredth.

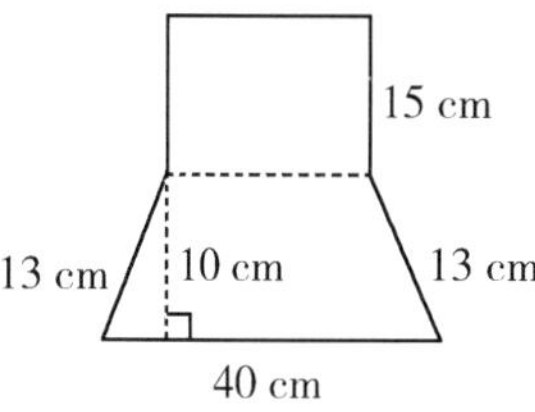

2. Find the perimeter and the area of the following sector if the central angle $\angle O$ is 120° and the radius is 15 cm. Round the answers to the nearest hundredth.

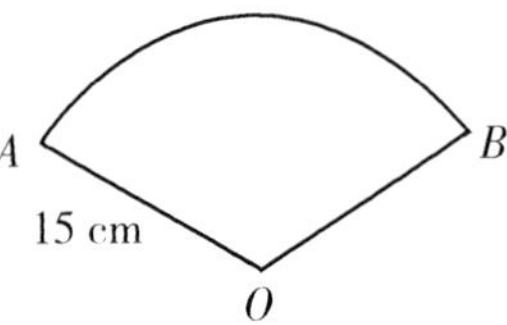

3. Find the perimeter and the area of the following figure and round the answers to the nearest hundredth.

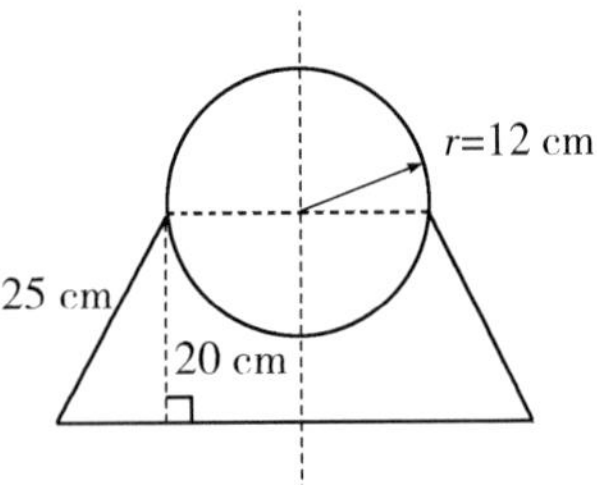

■Level B

1. The area of a square is $8.28\ m^2$. What is the area of a circle with the same perimeter as the square? Round your answer to the nearest hundredth.

2. The area of a sector is $\frac{4}{3}\pi$ and the central angle is 120°. Find the radius of the sector.

3. The area of a circle is 16 cm^2. What is the area of a square with the same perimeter as the circumference of the circle? Round your answer to the nearest hundredth.

4. Bob cut a rectangular paper that measures 60 cm by 30 cm into two parts. The two parts can be combined to form a triangle, a parallelogram or a trapezoid. Calculate the area of each shape that was cut.

5. The two parallel sides of a trapezoid measure 4 metres and 3 metres. If the height of the trapezoid is 2 metres, what is the area of the trapezoid?

6. The area of a parallelogram is 48 cm^2. If the height of the parallelogram is 4 cm, then what is the base of the parallelogram?

7. The area of a rectangle is 11 m^2. If the length of the rectangle is 5.5 m, then what is the perimeter of the rectangle?

8. The area of a triangle is 60 cm^2. If the base of the triangle is 8 cm, then what is the height of the triangle?

9. The perimeter of a rectangle is 60 metres. If the length of the rectangle is 16 metres, then what is the area of the rectangle?

10. The circumference of a circle is 62.8 cm. What is the diameter of the circle? (use pi=3.14)

11. The area of a triangle is 80 cm^2. If the base of the triangle is 8 cm, then what is the height of the triangle?

12. The area of a rectangle is 135 cm^2. If the width of the rectangle is 9 cm, then what is the perimeter of the rectangle?

13. The area of a triangle is 24 cm^2. If the base of the triangle is 4 cm, then what is the height of the triangle?

Level C

1. In the diagram below, an isosceles triangle is drawn inside a square and a circle is inscribed inside the triangle. One of the vertices(E) of the triangle is the midpoint of one side of the square. If the side length of the square is 5 cm, determine the length of $\overline{OD}$ to the nearest tenth.

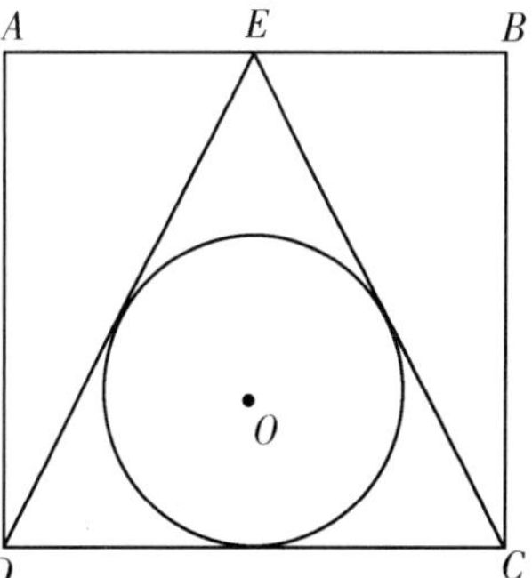

2. The equilateral triangle ABC with side lengths of 4 cm is drawn inside a circle (as shown in the diagram). Determine the radius of the circle to the nearest tenth.

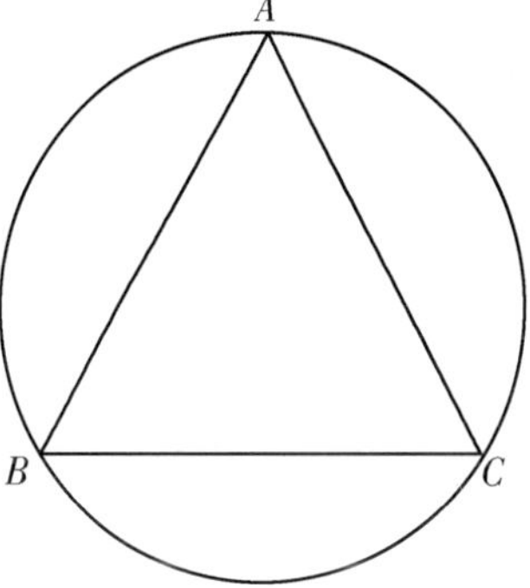

7.3 Tessellations

Regular Polygons

Regular Polygons are polygons whose side lengths and interior angles are all equal.

Sum of Interior Angles of a Polygon

The formula for calculating the sum of the interior angles of a polygon is:
$S=180°(n-2)$, where n is the number of sides

Derivation of Sum of Interior Angles of a Polygon

It is actually not difficult to understand the formula $S=180°(n-2)$. Let's use this hexagon as an example.

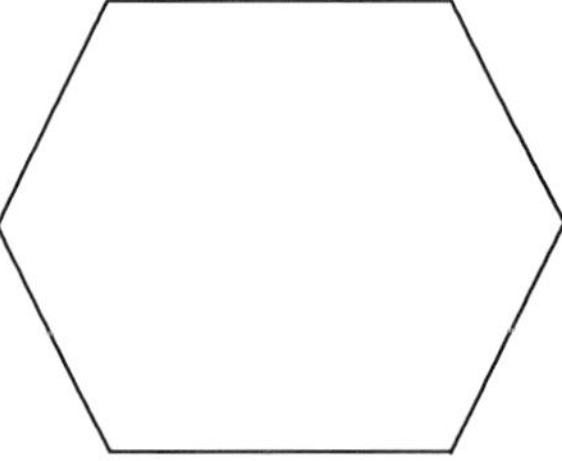

If we connect some of the vertices, we can obtain the following:

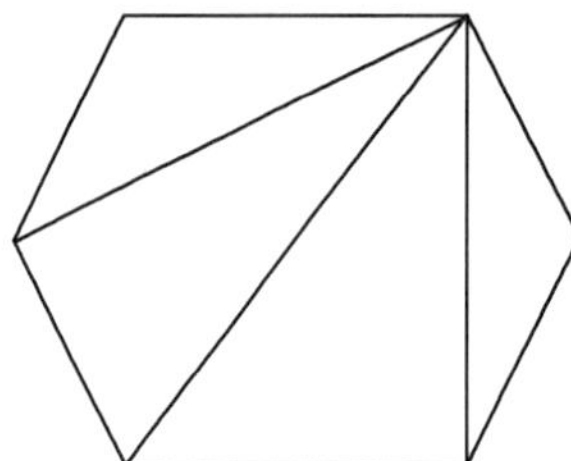

From the diagram, we can see that there are four separate triangles. We know that the sum of the interior angles in each triangle is equal to 180°. Since we have four triangles, the sum of the interior angles of the hexagon will be $180°\times4=720°$.
As you can see from the formula, we will always have $n-2$ triangles. The reason for this is because we simply cannot connect two adjacent vertices to make new triangles.

Example

Calculate the sum of the interior angles of a regular pentagon.

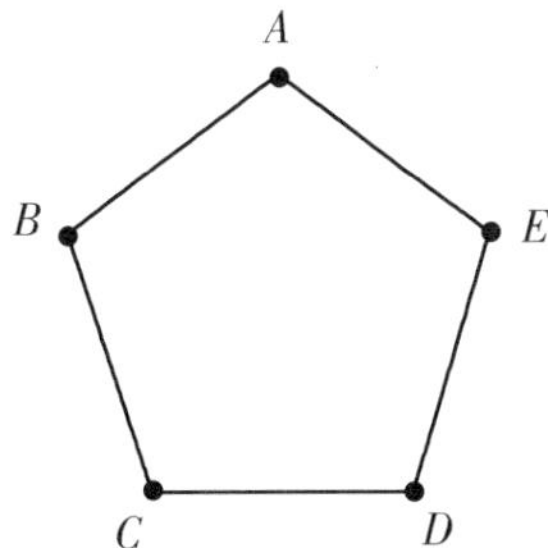

▶ *Solution*:

$S=180°(n-2)$, $S=180°(5-2)$, $S=3\times180°=540°$

Tiling Patterns

Tiling Patterns are patterns that repeat and fit together to completely cover a surface.

Tessellation

Tessellation is the tiling of a plane using one or more geometric shapes without overlaps and gaps. Tessellation consists of translations, rotations, and reflections.

Three Ways of Transformation to Make Tessellations

The **Translation** method to make tessellations is done by sliding a shape without rotating it.

The **Reflection** method to make tessellations is done by reflecting a shape over a mirror line. The line is a "line of symmetry" of the new shape.

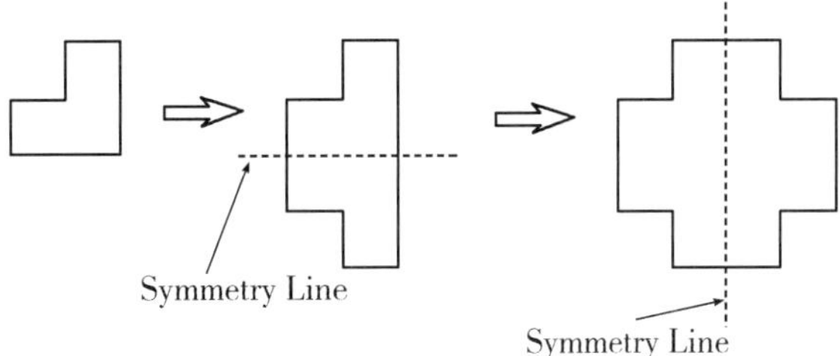

The **Rotation** method to make tessellations is done by rotating a shape around a rotational centre.

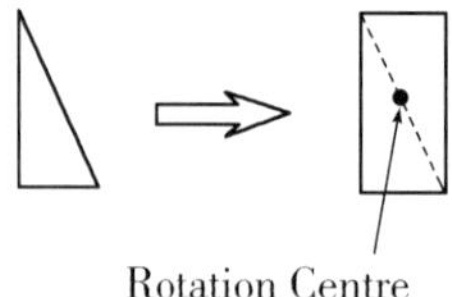

Practice

■Level A

Use the following shapes to answer the questions.

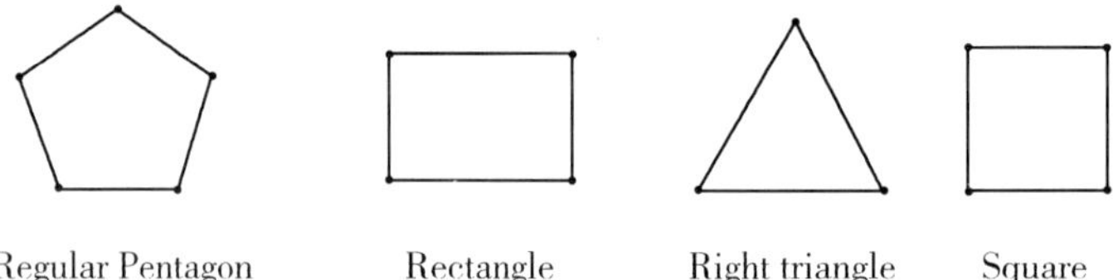

a) What is the sum of all the interior angles of each shape?

b) Find the measure of one interior angle in each shape?

c) Which shape cannot make a tessellation?

d) Use the Translation Method on a rectangle to make a tessellation.

e) Use the Reflection Method on a square to make a tessellation.

f) Use the Rotation Method on an equilateral triangle to make a tessellation.

■Level B

1. Use four rectangles and a square to make a tessellation in the shape of a square. Assume the area of the large square is nine times the area of the shaded square. If we need the area of the tessellation to be c^2, find $(a-b)^2$ in terms of c.

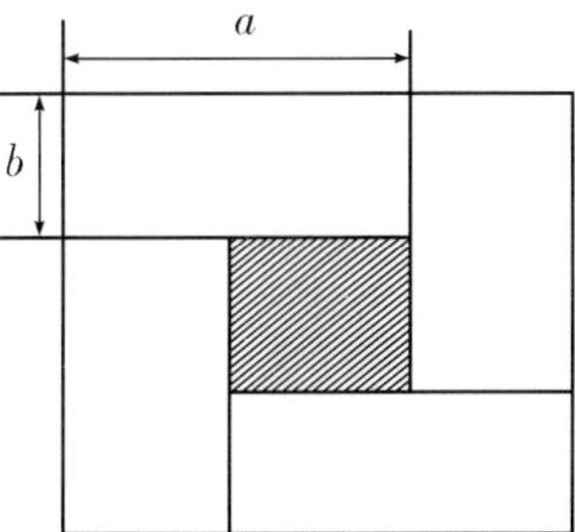

2. Eight of the same tiled rectangle form a big rectangle with a 40 cm side. What are the dimensions of each rectangle?

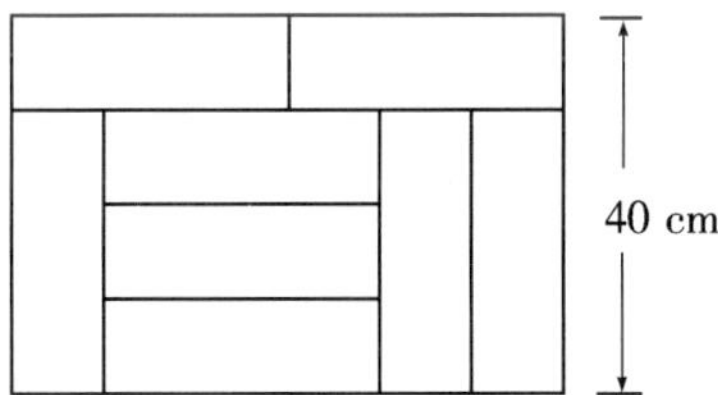

3. There are four kinds of floor bricks: a right triangle, a square, a regular hexagon, and a regular octagon. All of them have the same side length. Choose any two kinds of bricks to make a tessellation.

a) How many different combinations are there?

b) Draw a graph for each selection.

4. Which of these shapes can make a tessellation? Draw and colour your own tessellating pattern.

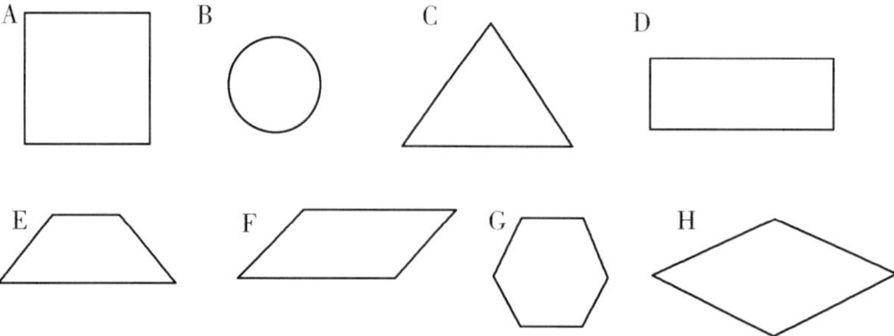

5.

Figure 1 Figure 2 Figure 3

a) Create Figure 3 according to Figure 1 and Figure 2, using parts of the circle and lines.

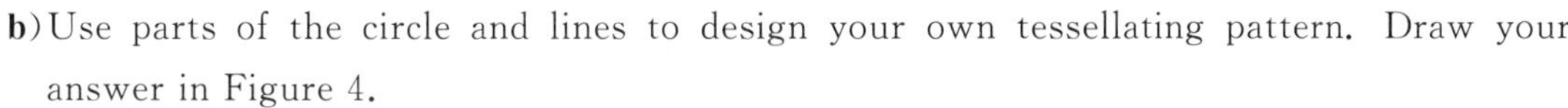

b) Use parts of the circle and lines to design your own tessellating pattern. Draw your answer in Figure 4.

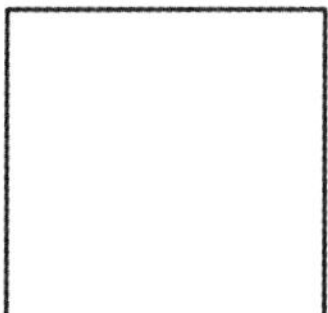

Figure 4

6.

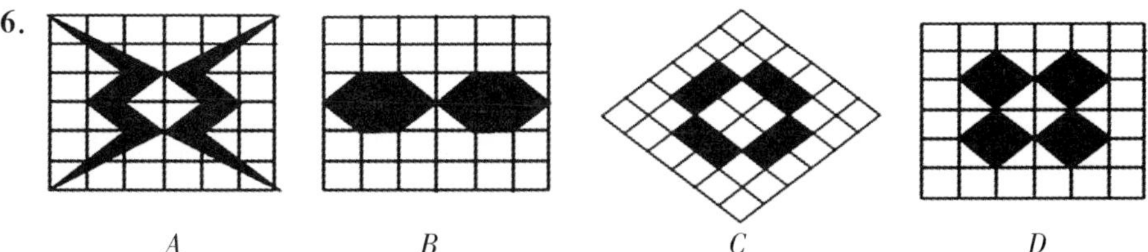

A *B* *C* *D*

Figures *A*, *B*, *C* and *D* were made up by translation and reflection from a basic graph. Find out what basic shape was used for each figure.

■Level C

1. Dina was calculating the sum of the interior angles of a regular polygon. She missed one interior angle, so her answer was 1999°. What is the angle that she missed?

2. Judy has three pieces of regular polygon-shaped cardboard. She wants to stick them together (as the diagram shows below). What is the shape of the third polygon if it is to touch the first two without a gap?

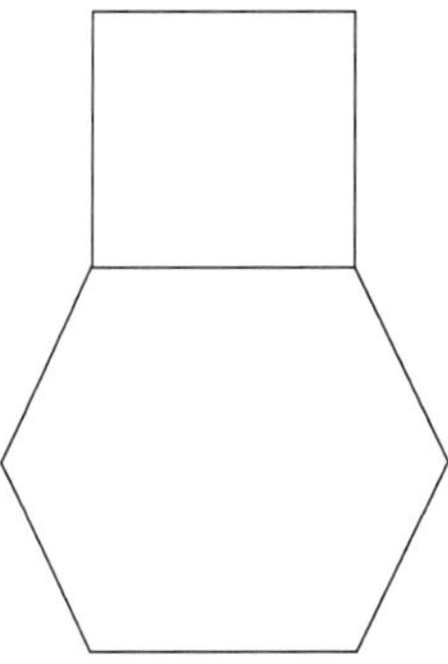

Chapter 7 TEST

1. Find the height AD, perimeter and area of $\triangle ABC$.

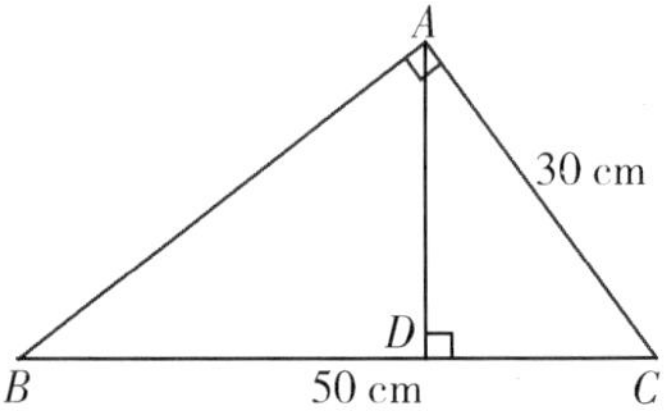

2. Find the length of Side y in the following figure.

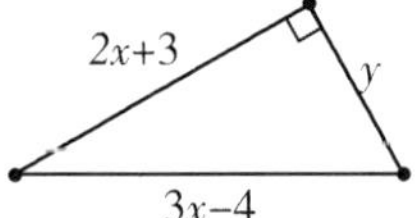

3. In the following triangle, $AB=AC=BC=30$ cm, D is the middle point of BC, and DE is perpendicular to AB. How long is DE?

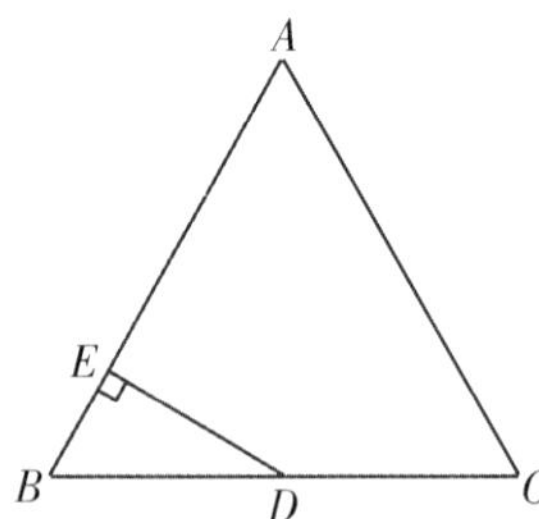

4. A rectangular paper $ABCD$ is folded on BD, so that Point C goes up to C'. If $AD=20$, and $AB=5$, how long is DE? Round your answer in the nearest hundredth.

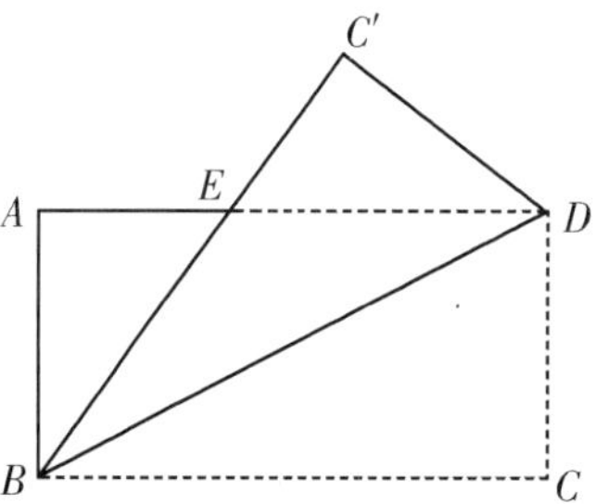

5. Find the perimeter and the area of the following sector if the central angle ($\angle O$) is $150°$ and the radius is 30 cm. Round your answers to the nearest hundredth.

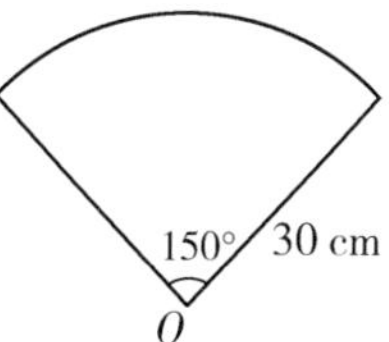

6. Find the perimeter and the area of the following figure. Round your answers to the nearest hundredth.

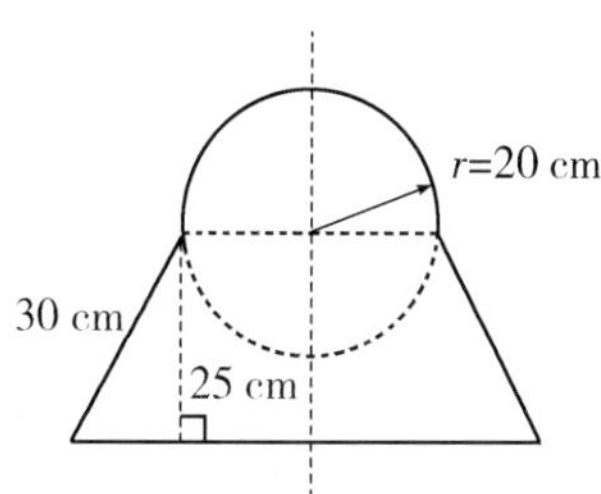

7. The area of a square is 100 m^2. What is the area of a circle with the same perimeter? (Use pi=3.14 and round your answer to the nearest hundredth if necessary.)

8. The area of a sector is $\frac{4}{3}\pi$ and the central angle is $120°$. Find the radius of the sector. (Leave your answer in exact form.)

9. The area of a circle is 314 cm^2. What is the area of a square with the same perimeter as the circumference of the circle? (Use pi = 3.14 and round your answer to the nearest hundredth.)

10. Allen cut a rectangular paper that measures 18 cm by 9 cm into two parts. The two parts can be combined to form a triangle, a parallelogram or a trapezoid. Calculate the area of each shape that was cut.

11. The 2 parallel sides of a trapezoid measure 16 cm and 12 cm. If the height of the trapezoid is 10 cm, what is the area of the trapezoid?

12. The area of a parallelogram is 24 cm^2. If the height of the parallelogram is 6 cm, then what is the base of the parallelogram?

13. The area of a rectangle is 30 m^2. If the length of the rectangle is 7.5 metres, then what is the perimeter of the rectangle?

14. The area of a triangle is 160 cm^2. If the base of the triangle is 18 cm, then what is the height of the triangle? (Round your answer to the nearest hundredth.)

15. The perimeter of a rectangle is 120 metres. If the length of the rectangle is 36 metres, then what is the area of the rectangle?

16. The area of a parallelogram is 48 cm^2. If the base of the parallelogram is 12 cm and one of angles of the parallelogram is 60°. What is the perimeter of the parallelogram? Round your answer to the nearest hundredth.

17. Figures 1 to 4 are made up by transformations (translations, reflections and rotations) of a basic shape.

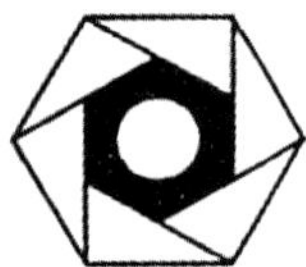
Figure 1

Figure 2

Figure 3

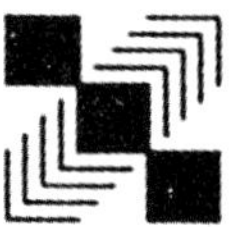
Figure 4

a) Find out the basic shape for each figure.

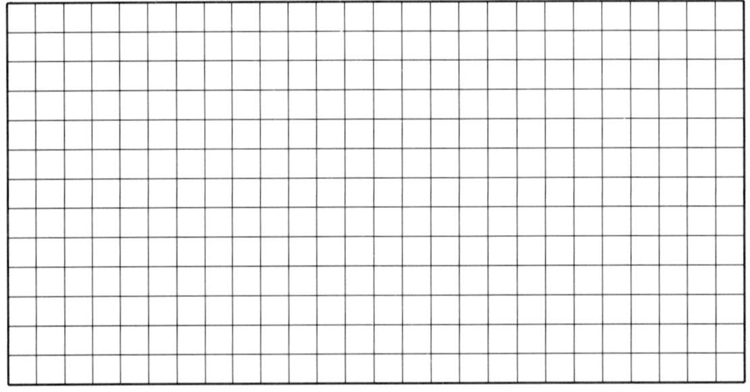

b) Use the basic shape to make up tessellating patterns different from figures 1 to 4.

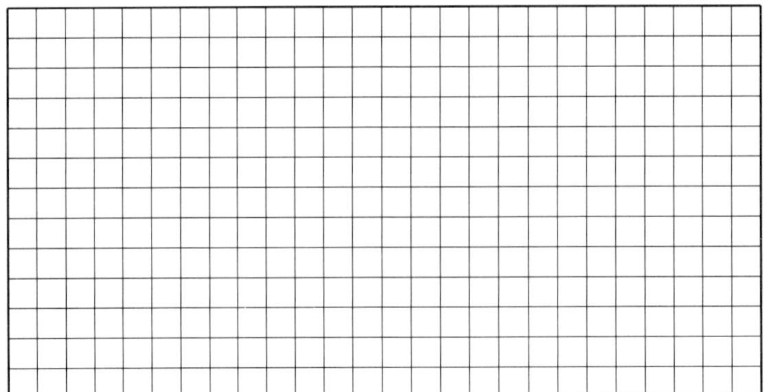

18. A quarter of a circle with a radius of 5 cm is shown below. Determine the area of the shaded region. Give an exact value.

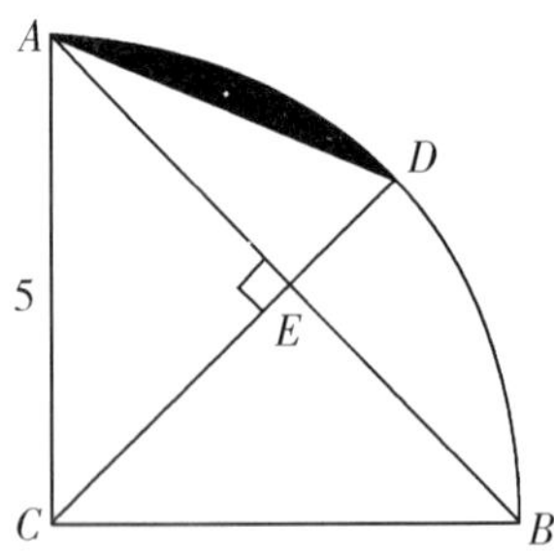

19. A circle is inscribed in a square with a side length of 7 cm. Triangle EGH is an equilateral triangle where Vertex E is the centre of the circle. What is the area of the shaded region? Round your answer to the nearest tenth.

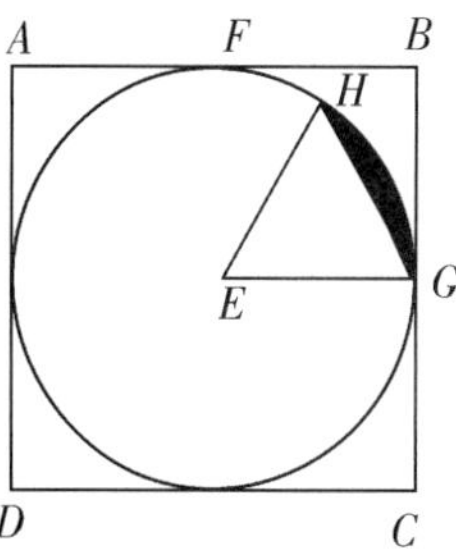

20. Triangle ABC is similar to Triangle DEF and E is the centre of the circle. Triangle ABC is an isosceles right-triangle with legs 11 cm long. Determine the length of $\overline{BG}$. Round your answer to the nearest hundredth.

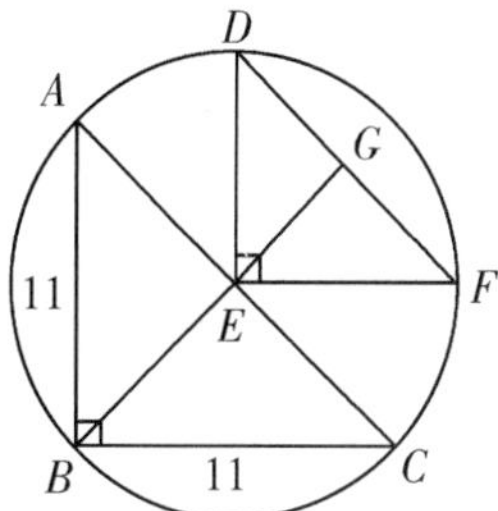

21. See the diagram below. If $\angle D : \angle E : \angle F = 4 : 5 : 6$, what is the ratio of $\angle A : \angle B : \angle C$?

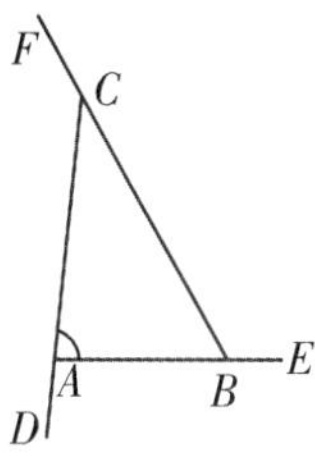

CHAPTER 8

Nets, Surface Area and Volume

8. 1 Nets

8. 2 Surface Area

8. 3 Volume

Chapter 8 TEST

Chapter 8 Nets, Surface Area and Volume

8.1 Nets

Net of an object

A **Net** is a two-dimensional pattern that when folded together produces a three-dimensional figure.

Example 1

Draw a net of the following rectangular prism.

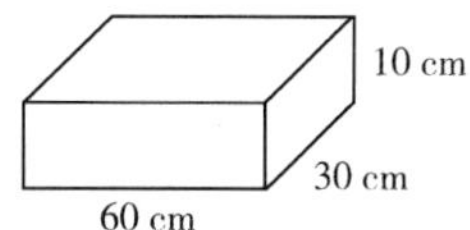

▶ *Solution*: See the net below.

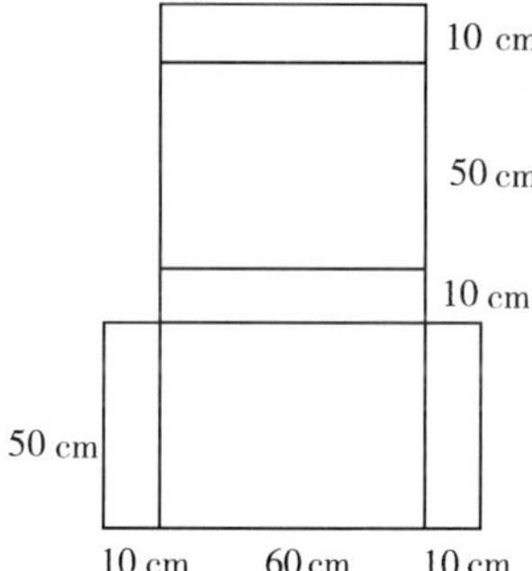

Example 2

Draw a net of the following cylinder.

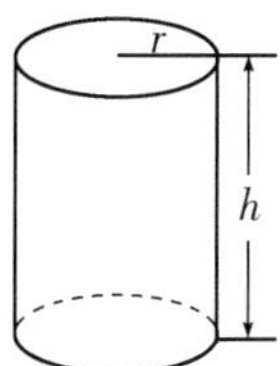

▶ *Solution*: See the net below.

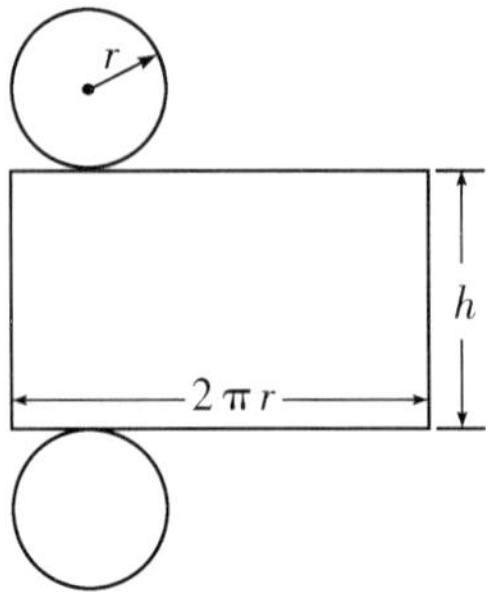

Example 3

Draw a net of the regular hexagonal prism. $a=3$ cm and $h=5$ cm.

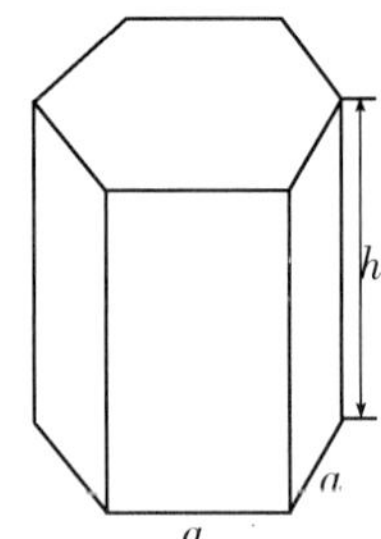

▶ *Solution*: See the net below.

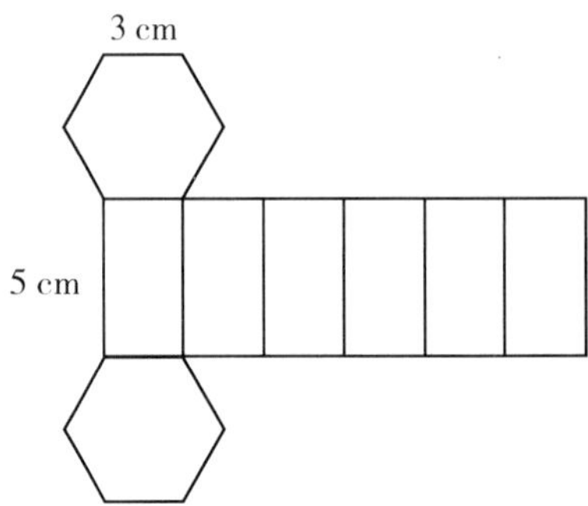

Example 4

Draw a net of the cone. $d=60$ cm and $h=40$ cm.

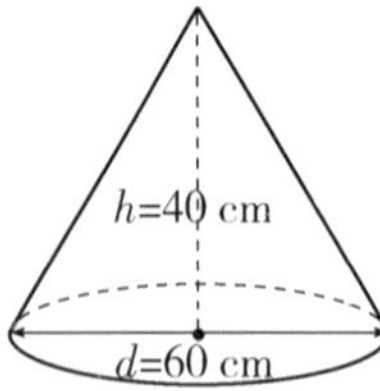

▶ *Solution*:

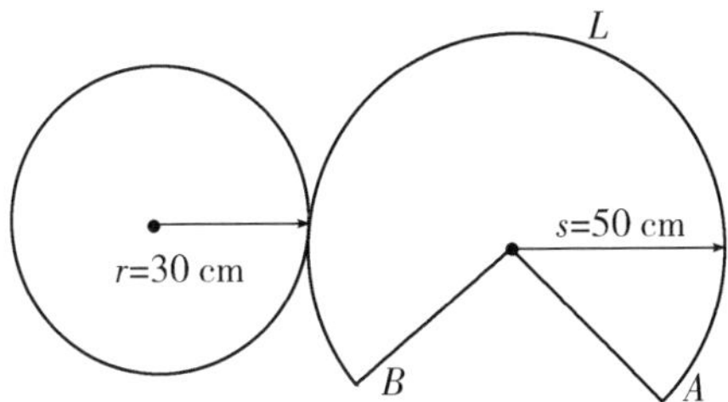

The net is a sector and a circle. The radius of the sector is s, and the arc length of the sector (between A and B) is L. The circumference of the base circle equals to L.

$r=\frac{d}{2}=\frac{60}{2}=30$ cm

$s=\sqrt{h^2+r^2}=\sqrt{40^2+30^2}=50$ cm

$L=2\pi r=\pi d=60\pi$ cm

Practice

■Level A

1. Find the object with the following net.

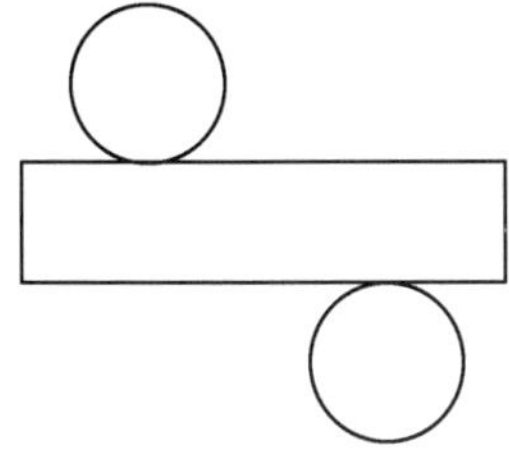

2. Draw a net of a cube with a side length of 5 cm.

3. Which of the following nets cannot be folded into a triangular prism?

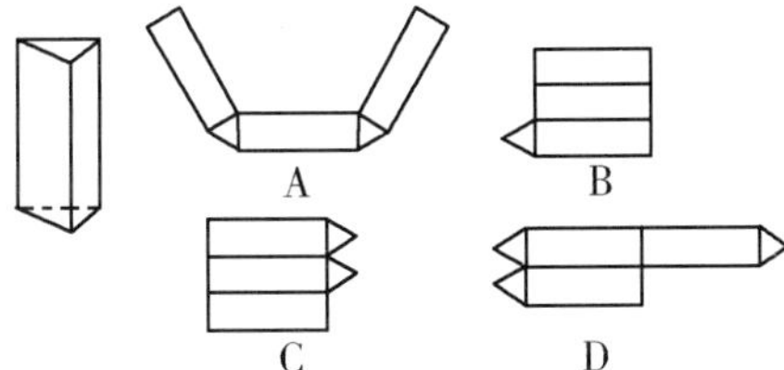

Level B

1. Which kind of object does the following net become after folding it together?

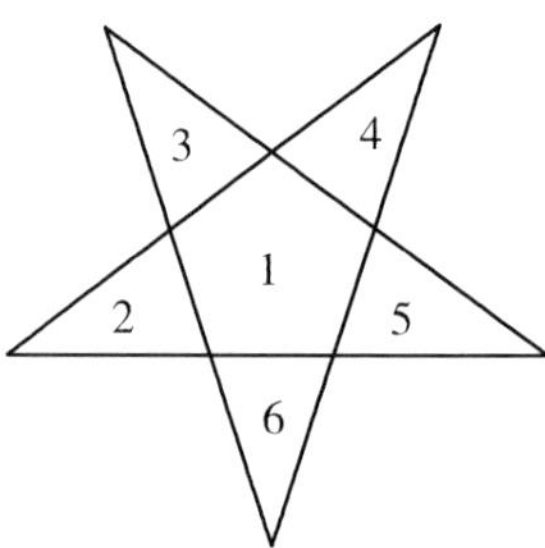

2. Draw a net for each object.

A

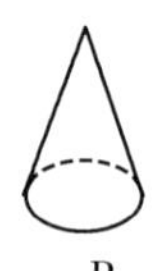
B

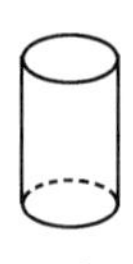
C

D

A.

B.

C.

D.

3. Draw a net of the following regular hexagonal prism.

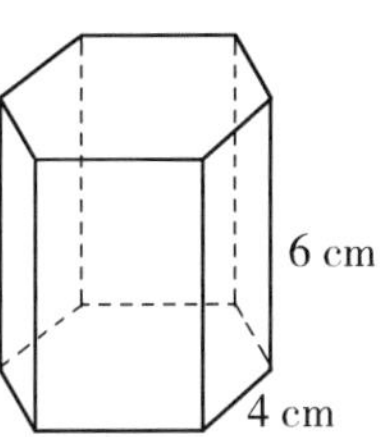

4. Draw a net of the following cone.

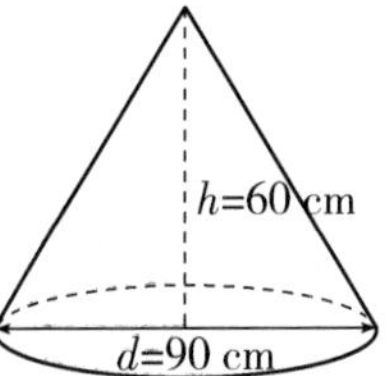

5. Draw the net of the following right hexagonal pyramid. The length of each side of the base is 3 cm. The height GH of the pyramid is 4 cm.

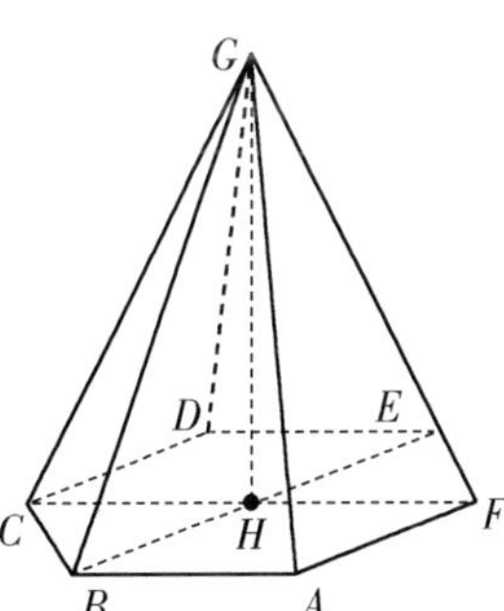

■Level C

1. A cube-shaped cardboard box is shown below. Draw a two-dimensional diagram of an unfolded box (Answers may vary).

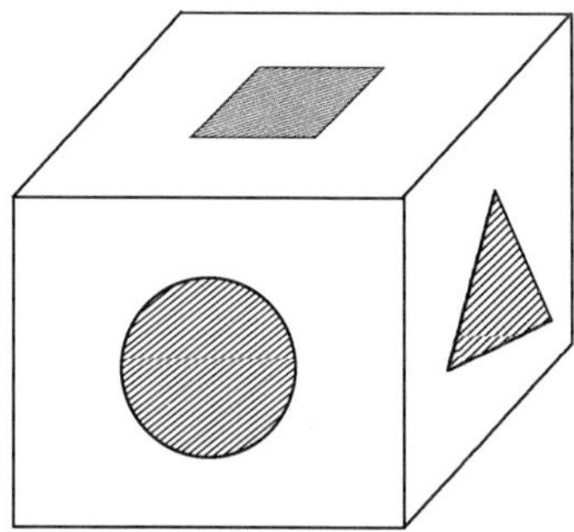

2. An unfolded cube box is shown below. Every side is labeled with a number. Every vertex is formed where three faces, and therefore three numbers, meet. If we multiply all three numbers from any given vertex, what is the greatest product possible?

	1		
4	2	5	6
	3		

3. A cube-shaped box is shown below. Point A and point B are labeled on two different vertices. What is the shortest distance between A and B. Give exact value.

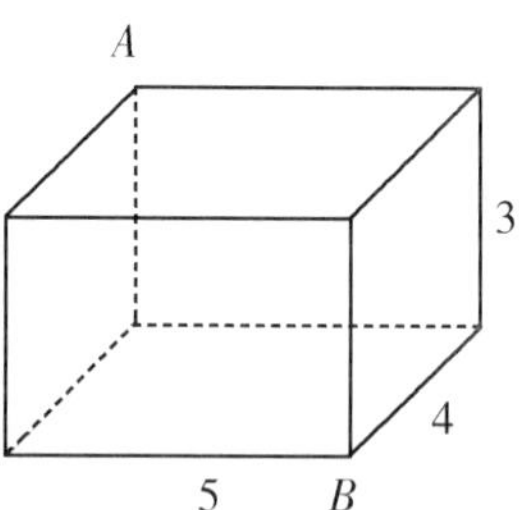

8.2 Surface Area

Surface Area

Surface Area is the sum of all face areas on a 3D object.

Basic steps to calculate Surface Area

1. Identify how many surfaces the object has.
2. Calculate the area of each surface.
3. Add all the areas of the surfaces together.

Example 1

Find the surface area for the rectangular prism.

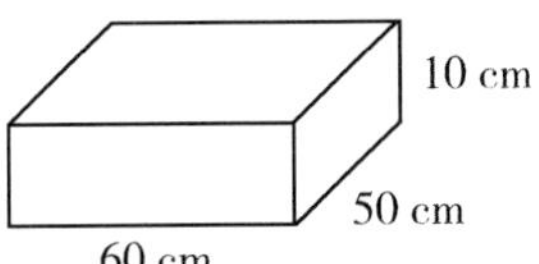

▶ *Solution*: The rectangular prism has 6 surfaces. The area of 3 different kinds of faces are calculated and added together. Since the other 3 are identical, multiply this answer by 2. So, the area is $A=(50\times 60+10\times 50+10\times 60)\times 2=8200$.

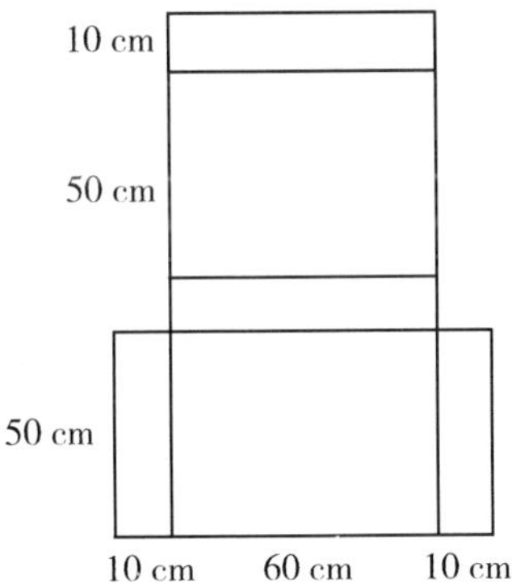

The surface area of the rectangular prism is 8200 cm^2.

Example 2

Find the surface area for the cylinder if $h = 10$ cm and $r = 5$ cm.

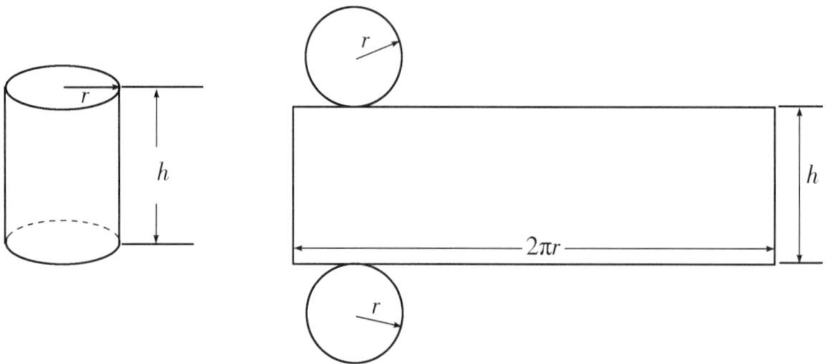

▶ *Solution*: The cylinder has 3 faces: two circles and a rectangle, and The dimensions are shown on the net.

So, the area $A = 2\pi rh + 2\pi r^2 = 2\pi r(h + r)$

$= 2 \times 3.14 \times 5(10 + 5) = 471 \text{ cm}^2$

The surface area of the cylinder is 471 cm^2.

Surface Area of Combined Shapes

Example 3

Calculate the surface area of the following composite object, where a triangular prism is placed on the top of a cube.

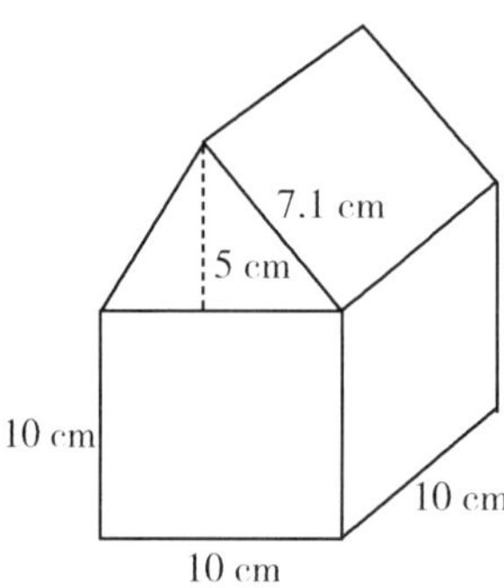

▶ *Solution*:

$A =$ Surface Area of Cube (only 5 faces) $+$ Area of 2 Triangles $+$ Area of 2 Rectangles

$A = 10 \text{ cm} \times 10 \text{ cm} \times 5 + 10 \text{ cm} \times 5 \text{ cm}/2 \times 2 + 10 \text{ cm} \times 7.1 \text{ cm} \times 2$

$= 500 \text{ cm}^2 + 50 \text{ cm}^2 + 142 \text{ cm}^2$

$= 692 \text{ cm}^2$

Example 4

Calculate the surface area for the following composite object. One rectangular block is sitting on the top of a larger rectangular block.

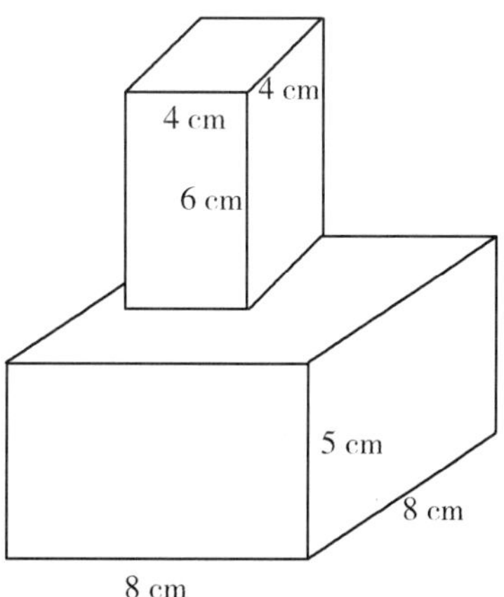

▶ *Solution*:

A = Bottom Area of Lower Block + Side Areas of Lower Block + Top Area of Lower Block + Side Areas of Upper Block + Top area of Upper Block

$= 8\text{ cm}\times 8\text{ cm} + 8\text{ cm}\times 5\text{ cm}\times 4 + (8\text{ cm}\times 8\text{ cm} - 4\text{ cm}\times 4\text{ cm}) + 4\text{ cm}\times 6\text{ cm}\times 4 + 4\text{ cm}\times 4\text{ cm}$

$= 64\text{ cm}^2 + 160\text{ cm}^2 + 48\text{ cm}^2 + 96\text{ cm}^2 + 16\text{ cm}^2$

$= 384\text{ cm}^2$

▶ *Better Solution*:

Note that from above, the top view looks like an 8 cm by 8 cm square.

A = Bottom Area + Top Area + Side Areas

= Area of Bottom view + Area of Top view + (Side Areas of Lower Block + Side Areas of Upper Block)

$= 64\text{ cm}^2 + 64\text{ cm}^2 + (160\text{ cm}^2 + 96\text{ cm}^2)$

$= 384\text{ cm}^2$

Practice

■Level A

1. Find the surface area of this figure.

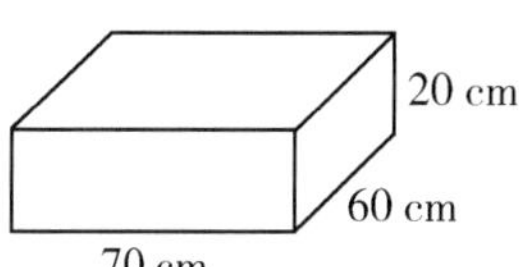

2. Find the surface area of this figure.

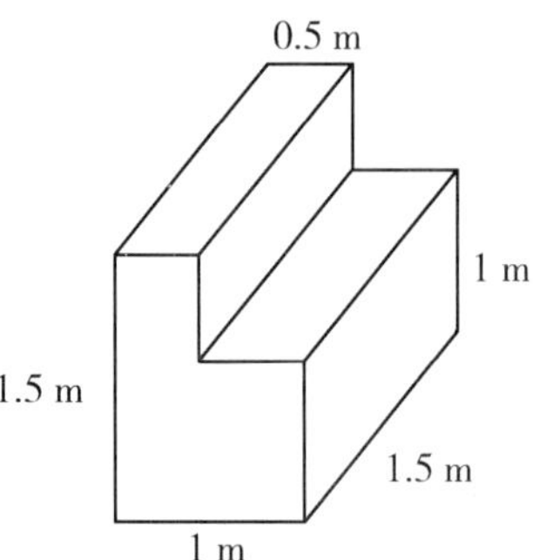

3. Find the surface area of the following figure if $h=20$ cm and $r=10$ cm.

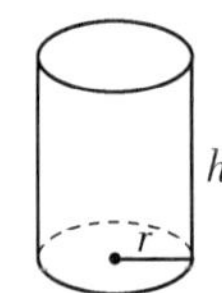

■Level B

1. A cylinder has a radius of 8 cm and a height of 12 cm. What is the surface area of this cylinder? Round your answer to the nearest cm^2.

2. A rectangular prism is combined with half a cylinder as shown in the following figure. Find the surface area.

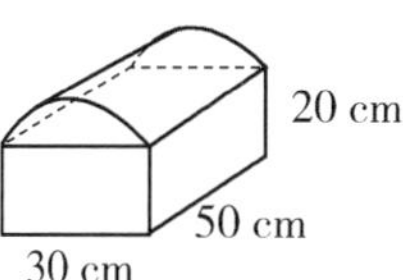

3. The dimensions of a room are 10 m × 8 m × 3 m. If the painting costs $\$15/m^2$, how much does it cost to paint the whole room? Round your answer to the nearest dollar.

4. The surface area of a rectangular prism is 600 cm^2. The length is 15 cm and the width is 10 cm. Find the height.

5. The surface area of a rectangular prism is 150 m^2. The area of the base is 15 m^2, and the length is 10 m. Find the width and height. Round to the nearest hundredth.

6. A cylinder has a height of 10 cm and a surface area of 78 cm^2. Find the radius and volume of this cylinder. Round to the nearest hundredth.

7. Find the surface area of the following figure. The radius of the cylinder is 0.5 m. Round to the nearest hundredth.

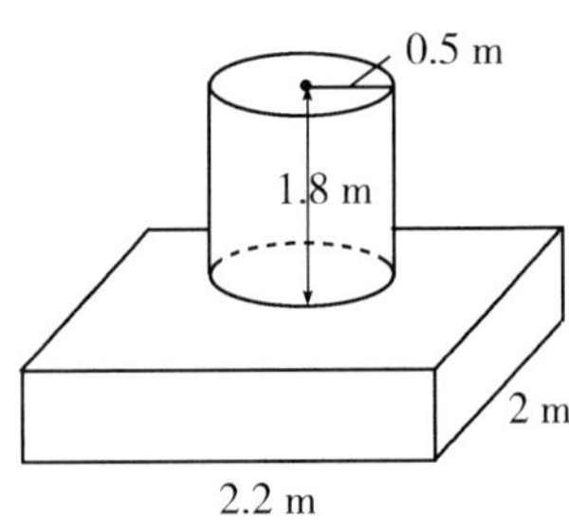

8. In a cylinder, $r=7$ cm and the surface area $S=665.89$ cm^2. Find the height of the cylinder. Round your answer to the nearest hundredth.

9. Find the surface area of the following figure. Round to the nearest hundredth.

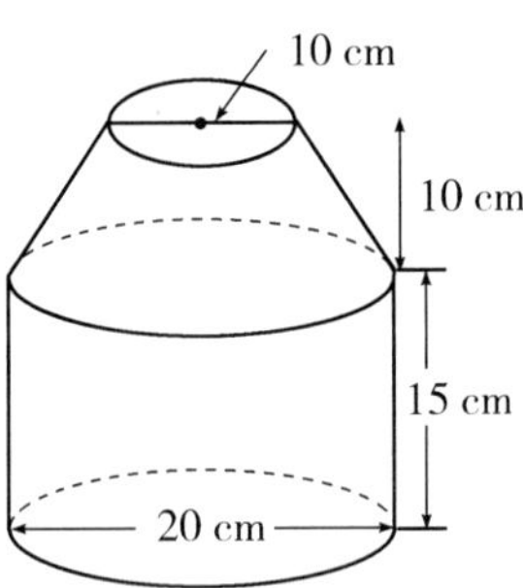

■Level C

1. A cube is placed on the top of a cylinder. Find the surface area of the combined shape. Round to the nearest tenth.

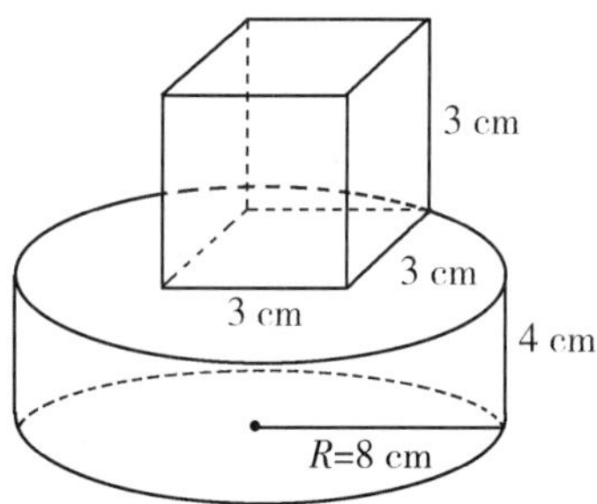

2. Find the surface area of the cylinder with a square hole in it. Round to the nearest tenth.

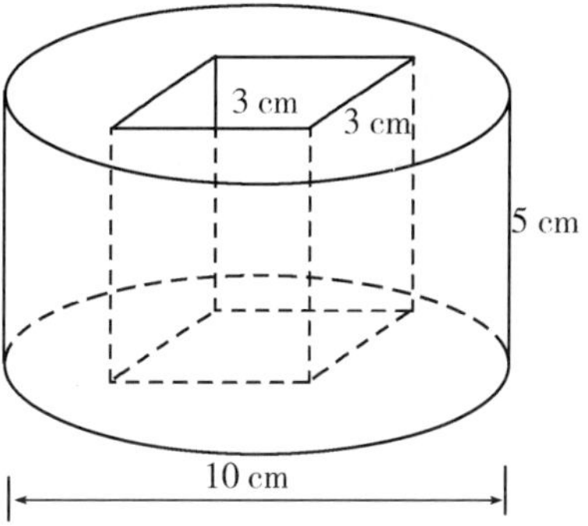

8.3 Volume

Definition of Volume

The **Volume** of a solid is how much space it takes up.

Definition of a Prism

A **Prism** is an object with a uniform cross section. A right prism is a prism that the top base is directly above the bottom base with all its lateral faces as rectangles. For example,

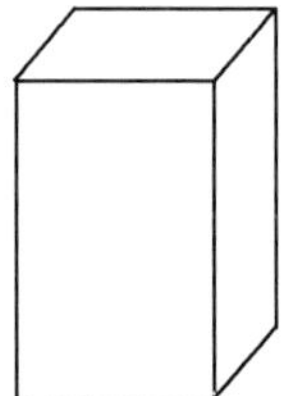

Basic Steps to Calculate the Volume of a Prism

1. Find the area of the base of the object.

2. Calculate the volume (V) using V=Base Area×H.

***H*:** **Height** is the perpendicular distance between the two bases.

Example 1

Find the volume of the rectangular prism.

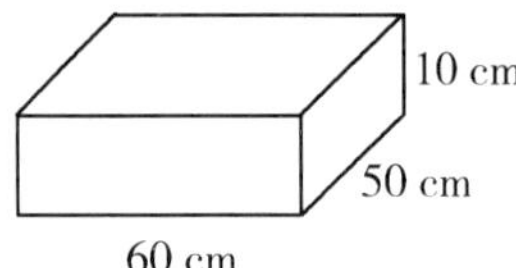

▶ *Solution*: The base area could be $50\times10=500$ cm^2. (or any one of 6 surfaces)

Volume V=base area$\times H=500\times60=30\ 000$ cm^3

(H: Height 60 is the perpendicular length to the base surface of the object.)

Example 2

Find the volume for the cylinder if $h=20$ cm and $r=10$ cm.

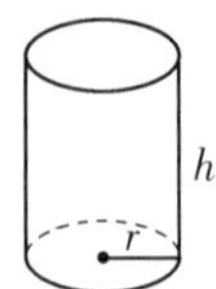

▶ *Solution*: The cylinder's base is the bottom circle or the top circle. It has a radius (r) of 10 cm.

So the volume is $V=\text{Base Area} \times h=\pi r^2 \cdot h=3.14\times 10^2\times 20=6280\ \text{cm}^3$.

The volume of the cylinder is $6280\ \text{cm}^3$.

Volume of Combined Shapes

Example 3

Calculate the volume of the following dumbbell and round to the nearest tenth.

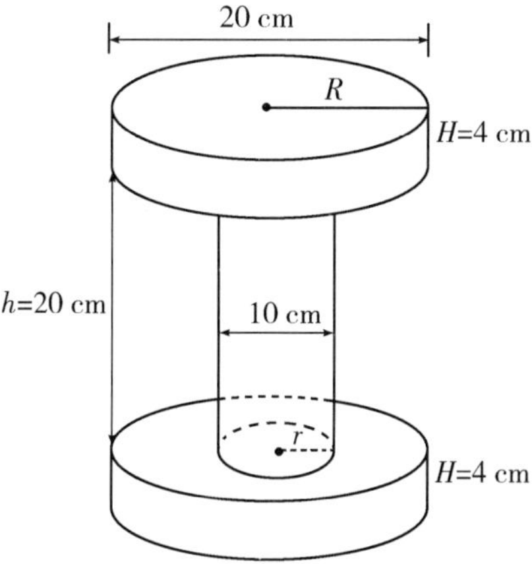

▶ *Solution*:

$V=\text{Volume of Middle Cylinder} + \text{Volume of Side Discs}\times 2$

$=\pi r^2 h + 2\pi R^2 \cdot H$

$=5\ \text{cm}\times 5\ \text{cm}\times \pi\times 20\ \text{cm}+10\ \text{cm}\times 10\ \text{cm}\times \pi\times 4\ \text{cm}\times 2$

$=500\pi\ \text{cm}^3+800\pi\ \text{cm}^3$

$=1300\pi\ \text{cm}^3$

$=4082.1\ \text{cm}^3$

Example 4

Calculate the volume of the following disc with a square hole in the middle. Round to the nearest tenth.

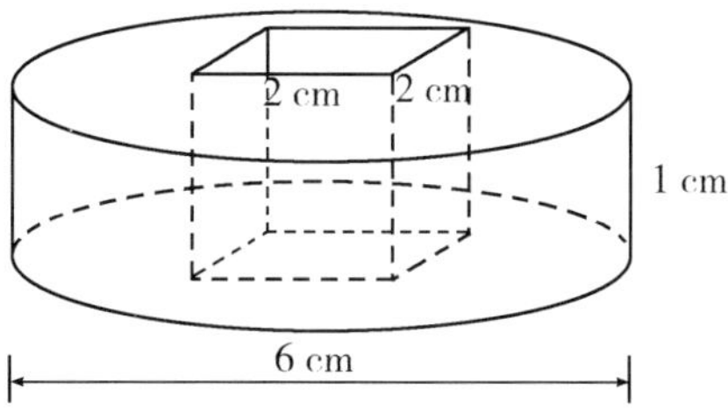

▶ *Solution*: V = Volume of Disc − Volume of Hole

$= 3\text{ cm} \times 3\text{ cm} \times \pi \times 1\text{ cm} - 2\text{ cm} \times 2\text{ cm} \times 1\text{ cm}$

$= 9\pi\text{ cm}^3 - 4\text{ cm}^3$

$= 24.3\text{ cm}^3$

Practice

■Level A

1. A cylinder has a radius of 2 m and a height of 3 m. What is the volume of this cylinder? Round to the nearest tenth.

2. The dimensions of a storage room are 20 m × 10 m × 3 m. If the volume of a box is 0.5 m^3, how many boxes can be stored in the room?

3. The volume of a rectangular prism is 64 m^3. The base area is 8 m^2. Find the height.

4. Find the volume of the following figure. Round to the nearest tenth.

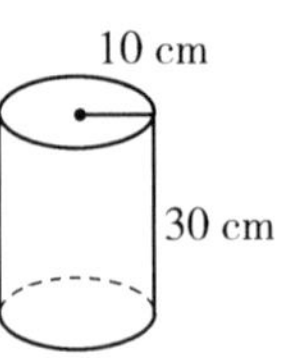

■Level B

1. Find the volume of this figure.

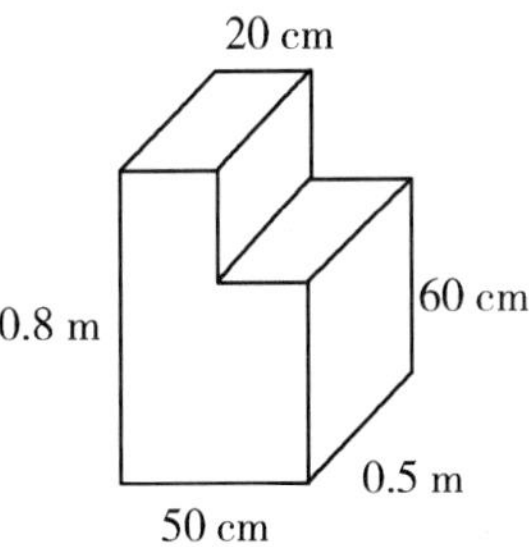

2. A rectangular prism is combined with half a cylinder as shown in the following figure. Find the volume. Round to the nearest hundredth.

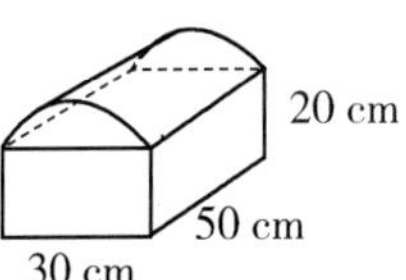

3. The volume of a rectangular prism is 210 m^3. The length is 7 m and the width is 6 m. Find the height.

4. The volume of a rectangular prism is 264 cm^3. The area of the base is 44 cm^2, and the length is 8 cm. Find the width and height. Round to the nearest tenth.

5. A cylinder has a height of 10 cm and a volume of 314 cm^3. Find the radius and the surface area of this cylinder. Round to the nearest hundredth.

6. Find the volume of the following figure. The radius of the cylinder is 0. 2 m. Round to the nearest hundredth of a cm^3.

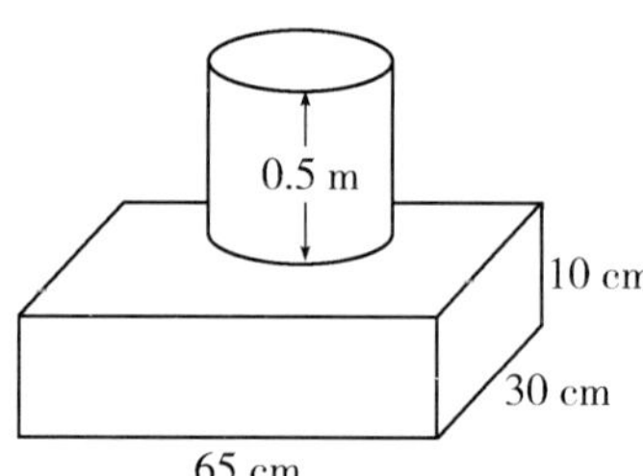

7. A cylinder has a radius of 15 cm and a volume of 9891 cm^3. Find the height and surface area of the cylinder. Round to the nearest hundredth.

8. Find the volume of the following figure.

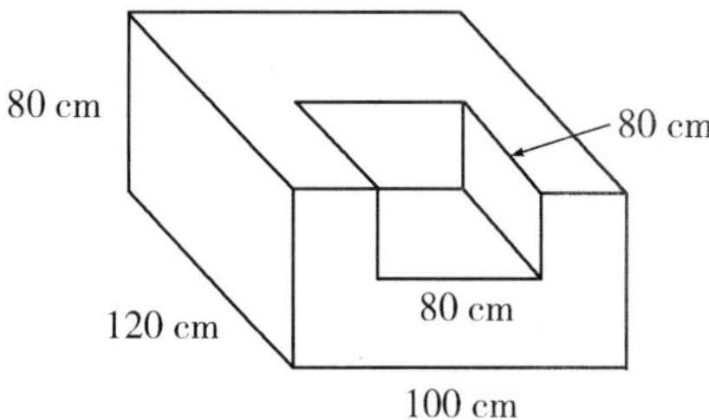

■Level C

1. A cube is placed on top of a cylinder. Find the volume of the combined shape and round to the nearest tenth.

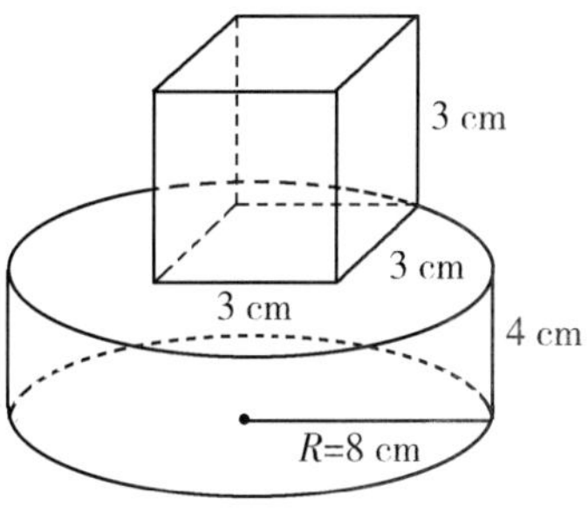

2. Find the volume of the cylinder with a square hole in it. Round to the nearest tenth.

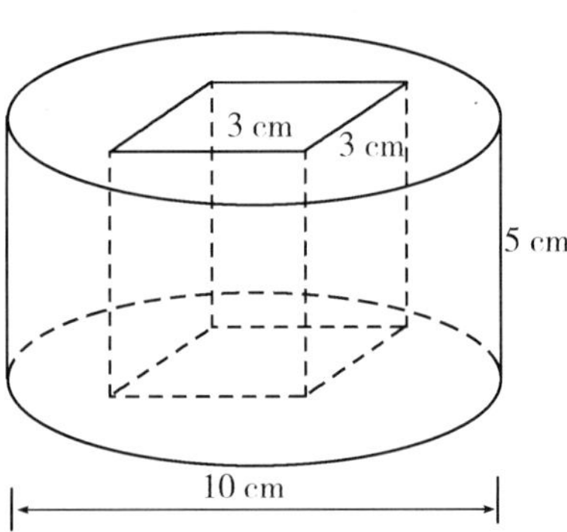

3. Find the volume of the following figure (Note: the volume of a cone is $\frac{1}{3}\pi r^2 h$). Round to the nearest hundredth.

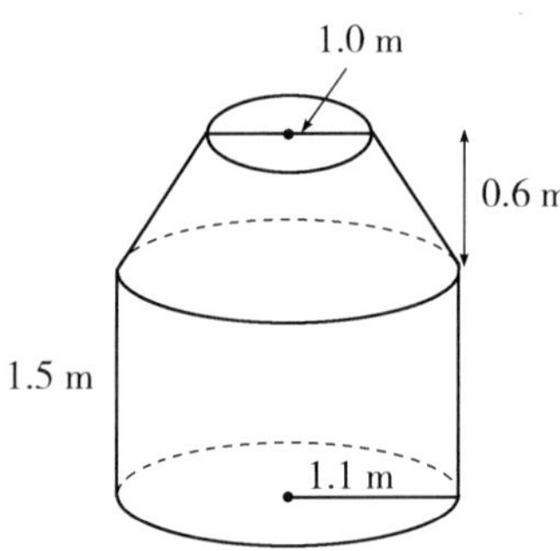

Chapter 8 TEST

1. A triangle's area is 8 m^2. The height of the triangle is 2 m. What is the length of the base of the triangle?

2. You have a triangular cooking pan at your camp. It makes great hot dogs on a camp fire. The triangle has an area of 10 m^2. The height is 2.5 m. What is the length of the base?

3. The dimensions of a warehouse are 30 m$\times$20 m$\times$5 m. If the volume of a case is 1.2 m^3, how many cases can be stored in the warehouse?

4. The volume of a rectangular prism is 128 m^3. The base area is 8 m^2. Find the height.

5. Find the surface area and the volume of the following cylinder. Give exact answers.

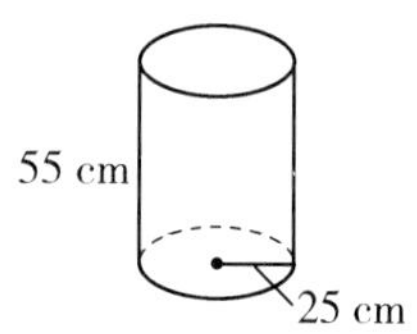

6. Based on the following regular hexagonal prism

a) Draw the net.

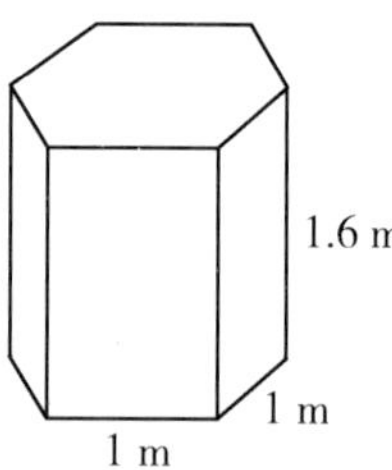

b) Find the surface area. Round to the nearest thousandth.

c) Find the volume. Round to the nearest thousandth.

7. Based on the following cone

a) Draw the net. Round to the nearest thousandth.

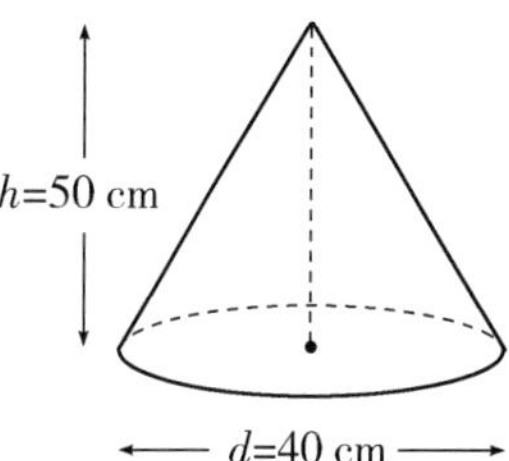

b) Find the surface area. Round to the nearest hundredth.

c) Find the volume. Round to the nearest hundredth.

8. Find the surface area and the volume for the following figure. Round to the nearest hundredth.

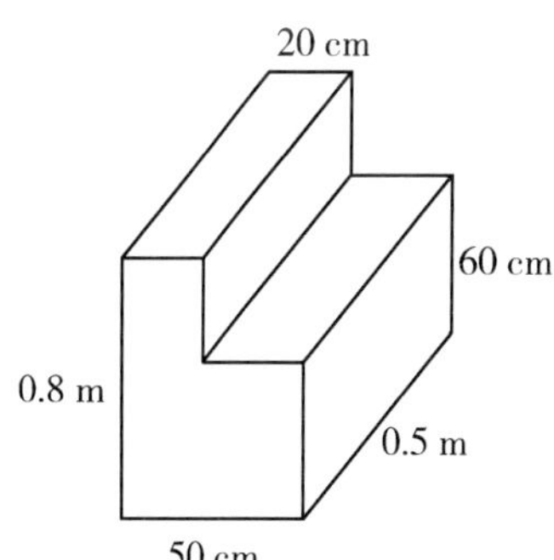

9. A rectangular prism is combined with half a cylinder as shown in the following figure. Find the surface area and the volume. Round to the nearest hundredth.

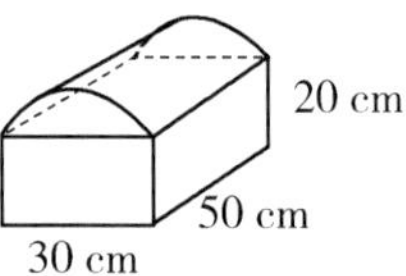

10. The volume of a rectangular prism is 24.99 m^3. The length is 3.5 m and the height is 3.4 m. Find the width. Round to the nearest hundredth.

11. The volume of a rectangular prism is 4284 cm^3. The area of the base is 357 cm^2, and the length is 21 cm. Find the width and height.

12. A cylinder has a height of 20 cm and a volume of 7598.8 cm^3. Find the radius and the surface area of this cylinder. Round to the nearest hundredth.

13. Find the surface area and the volume of the following figure. The radius of the cylinder is 20 cm. Round to the nearest hundredth.

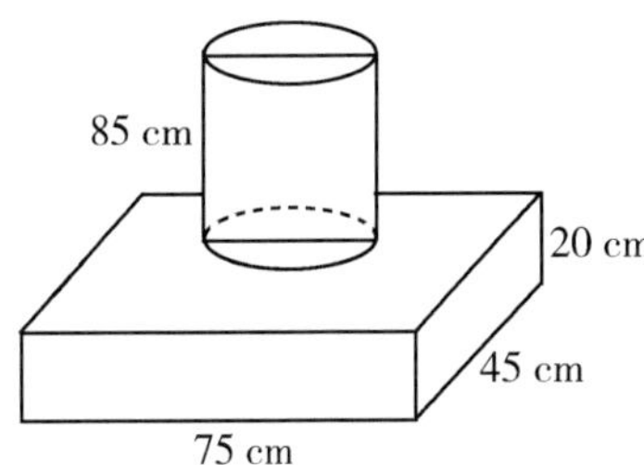

14. In a cylinder, $r=9$ cm and volume $V=25\ 434$ cm^3. Find the height and surface area of the cylinder. Round to the nearest hundredth.

15. Find the surface area and the volume of the following figure. Round to the nearest hundredth.

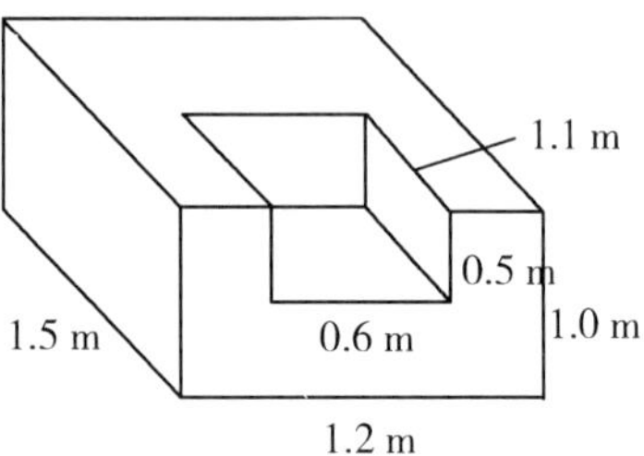

16. Find the surface area and the volume of the following figure. Round to the nearest hundredth.

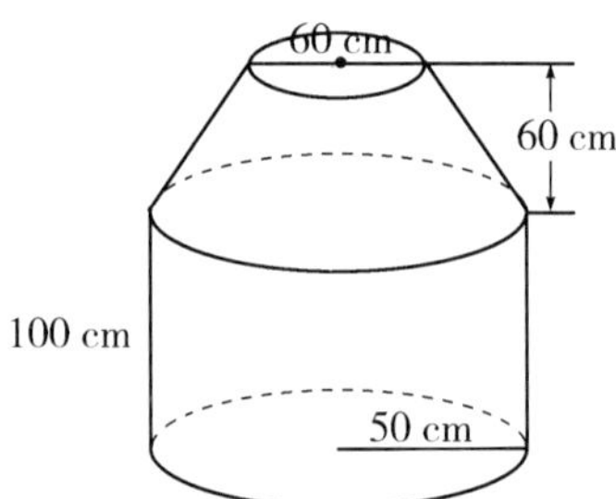

17. In the following right hexagonal pyramid, the side length of the hexagon is 5 cm. The height GH of the pyramid is 12 cm.

a) Find the surface area. Round to the nearest hundredth.

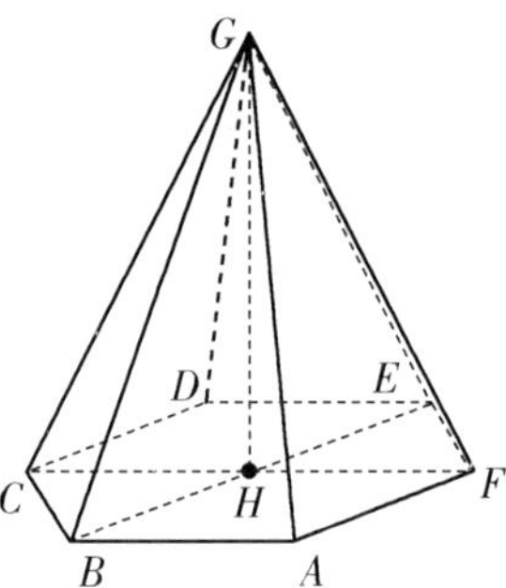

b) Find the volume. Round to the nearest hundredth.

18. Based on the diagram shown, state whether the following statements are true or false for the following questions.

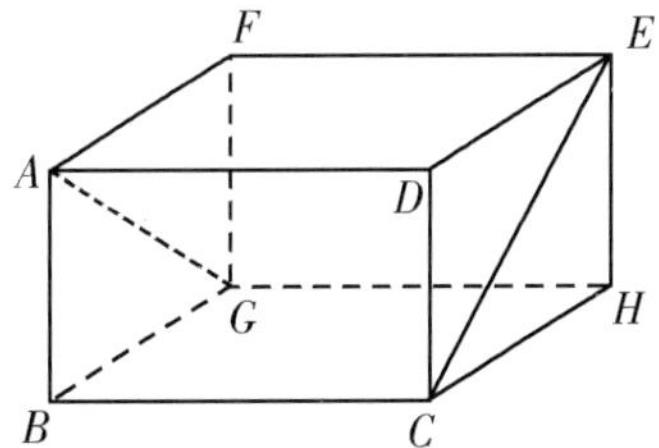

a) $\overline{AE}=\overline{CG}$ (T/F)

b) $\overline{AG}=\overline{CE}$ (T/F)

c) $\overline{BH}=\overline{FH}$ (T/F)

d) $\overline{AC}=\overline{EG}$ (T/F)

19. Draw the net of the rectangular prism which is shown below and label all vertices.

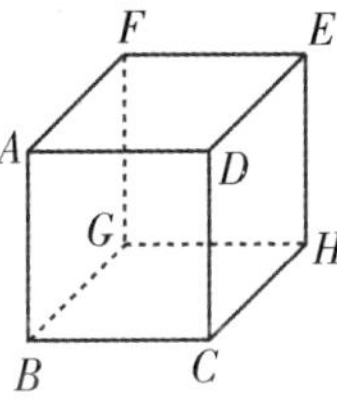

20. Which of the following nets are correct?

a)

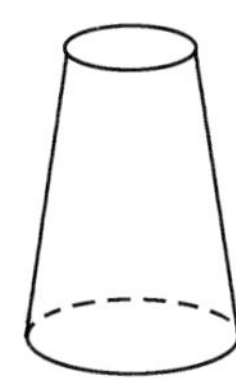

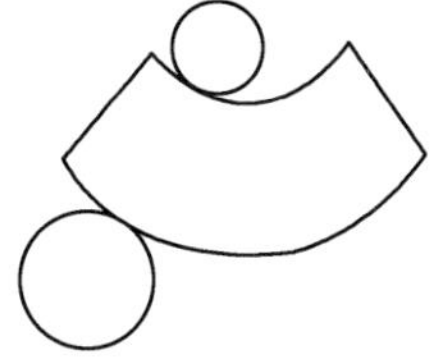

b)

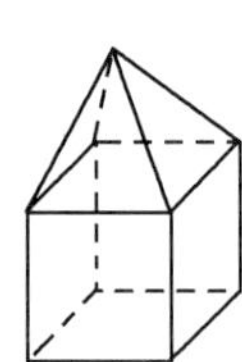

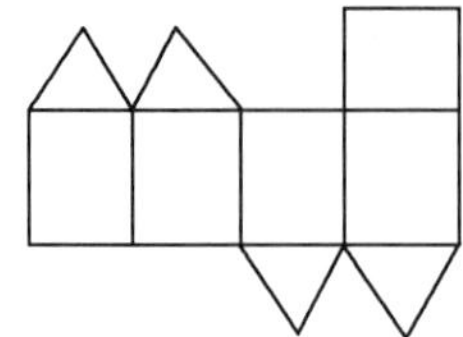

c)

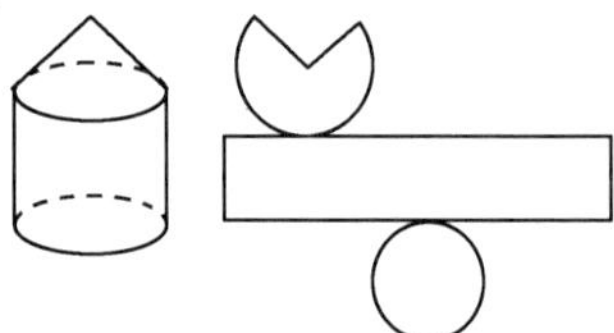

21. A cube is shown below. What is the measure of $\angle ABC$?

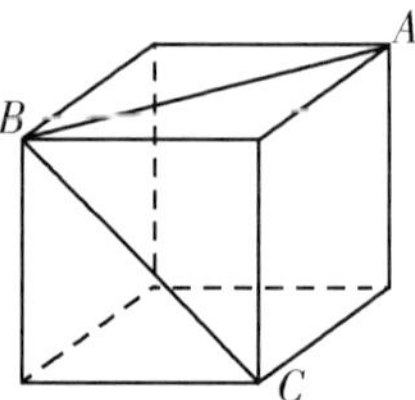

22. A cone is cut at the cross-section of $\overline{AB}$(see diagram below). If $\overline{PC}=\overline{PD}=18$, A & B are the midpoints of $\overline{PC}$ and $\overline{PD}$, $\overline{AB}=4$ and $\overline{CD}=8$, what is the surface area after the cone is cut? Give the exact value.

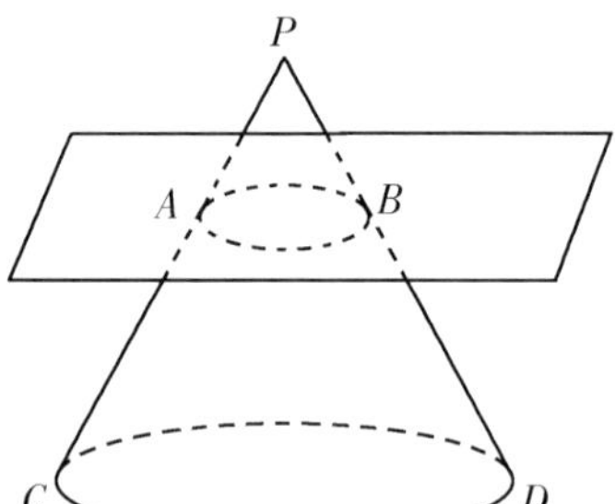

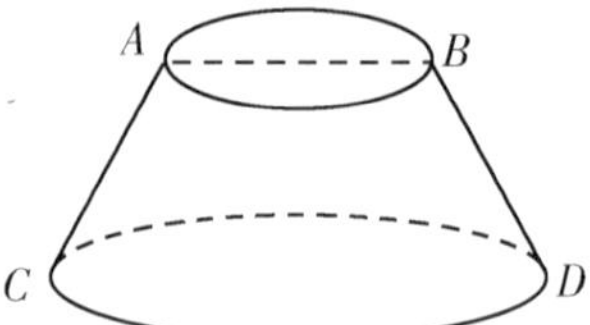

23. If you were to fold the following sectors, which sector will produce the highest cone?

a)

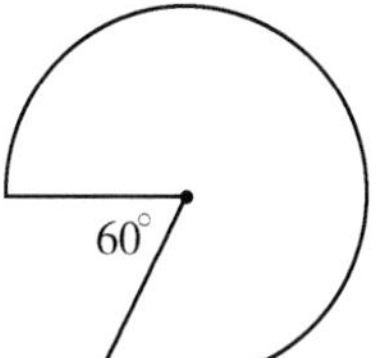

b)

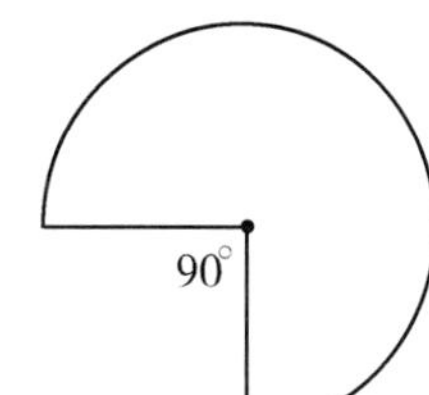

c)

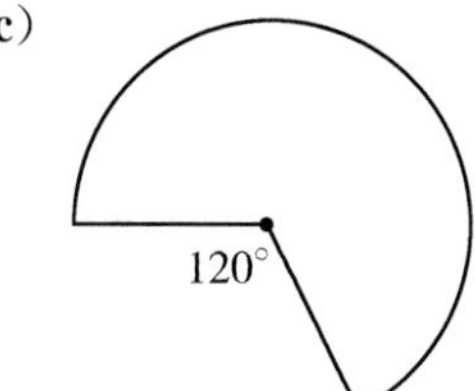

d)

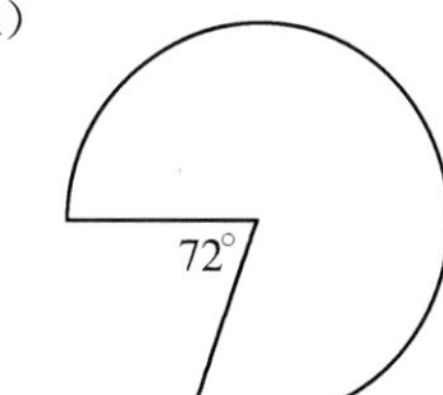

CHAPTER 9

3-D Views and Three Views of 3-D Objects

9. 1 Three Views from a 3-D View

9. 2 The 3-D View from Three Views of an Object

9. 3 3-D Views and Three Views of Combined Objects

Chapter 9 TEST

3-D Views and Three Views of 3-D Objects

9.1 Three Views from a 3-D View

To describe a three-dimensional object on a two-dimensional plane, we usually need to draw views from three different directions. We call these views as the **Front View, Side View and Top View.**

Example 1

Draw the front view, top view and side view for the object.

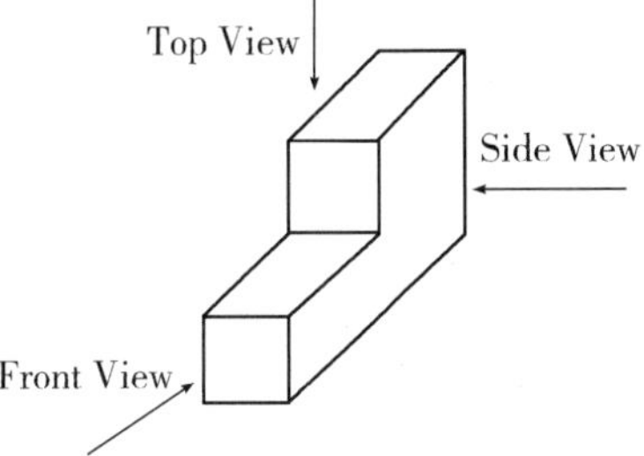

▶*Solution*:

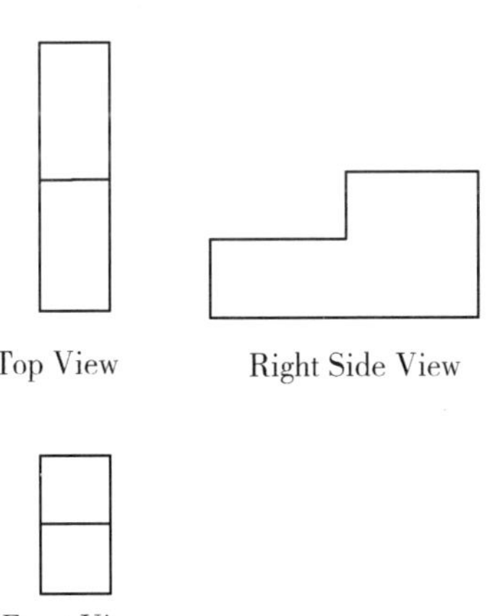

The front view should be below the top view, and the side view should be on the left side or right side of the top view depending on the viewing direction. In this example, we view the object from the right side, so the side view should be to the right of the top view.

The segments of the object's surfaces should be aligned in all three views.

Example 2

Draw the front view, top view and side view for the object.

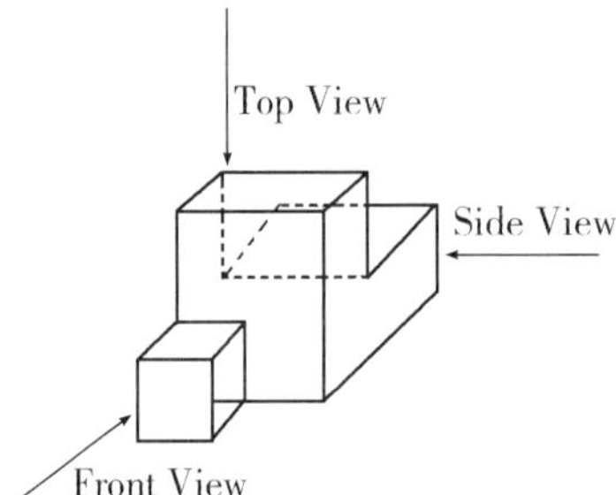

▶ *Solution*:

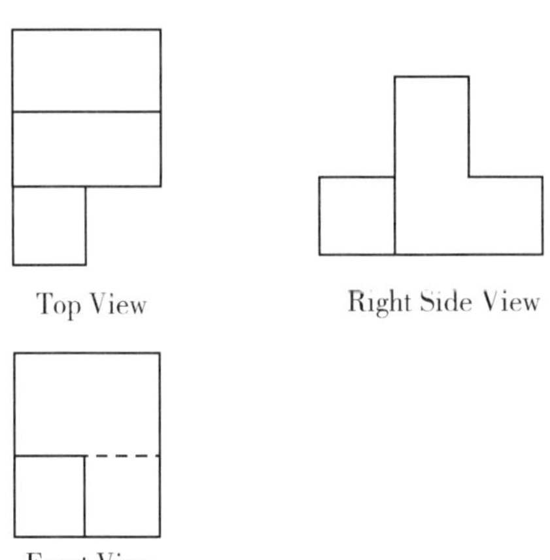

Practice

■Level A

1. Draw the front view, top view and right side view for the object.

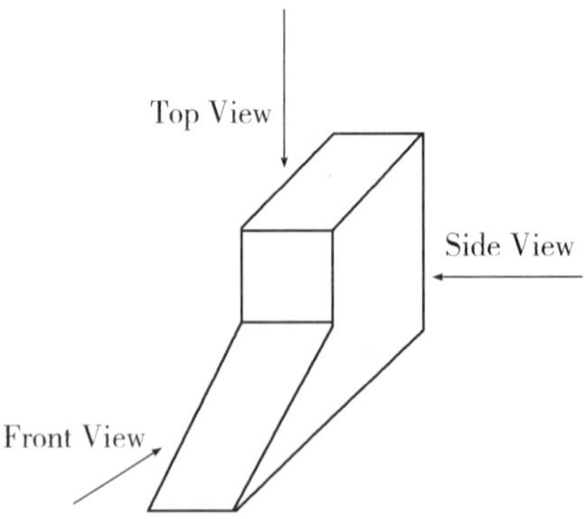

2. Draw the front view, top view and right side view for the object.

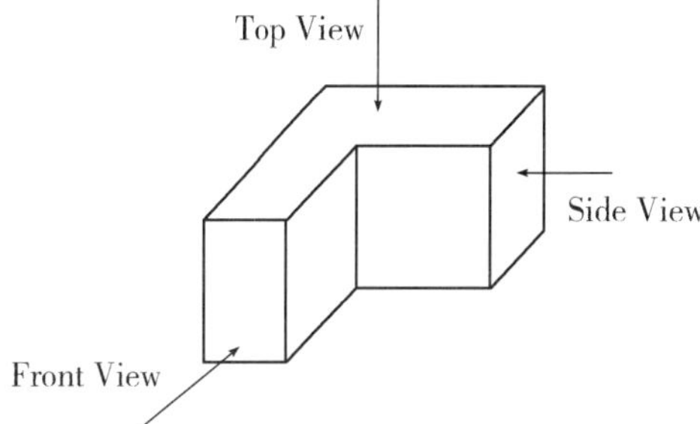

Level B

1. Draw the front view, top view and right side view for the object.

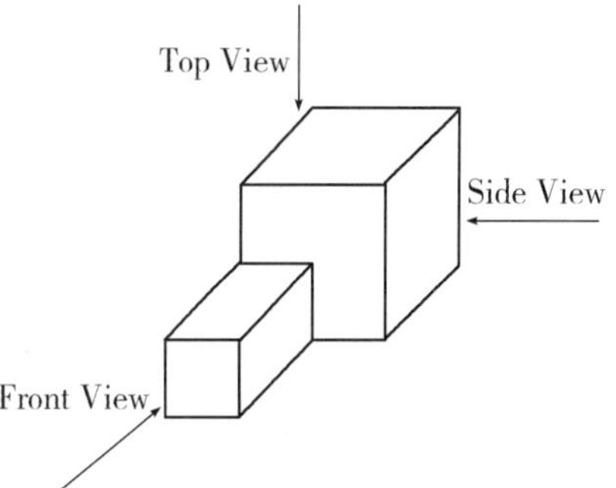

2. Draw the front view, top view and right side view for the object.

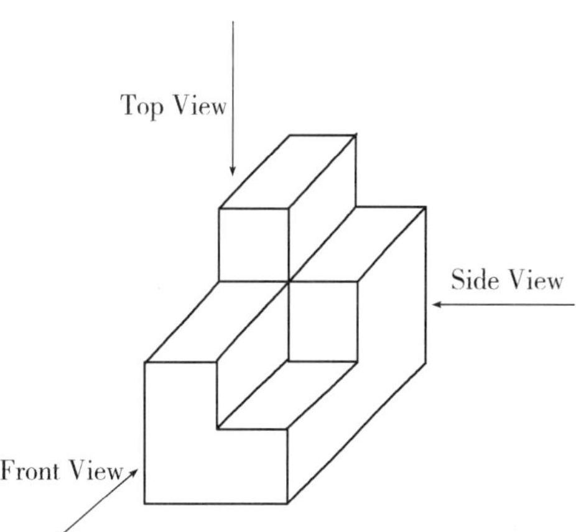

3. Draw the front view, top view and right side view for the object.

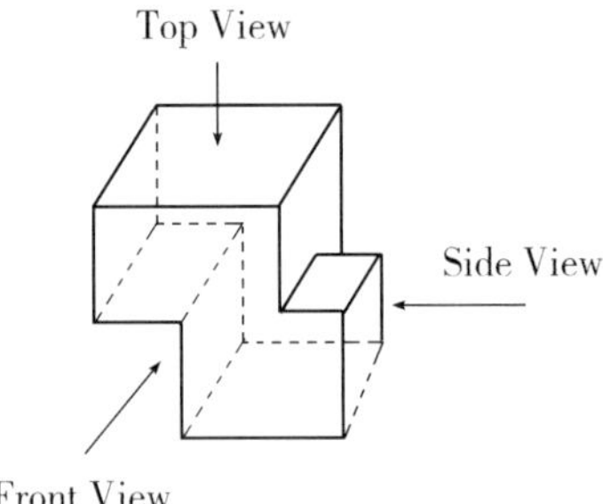

■Level C

1. Draw the front view, top view and right side view for the object.

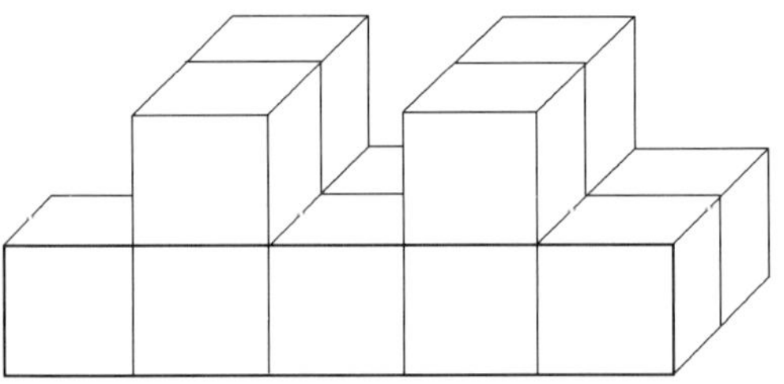

2. Draw the front view, top view and right side view for the object.

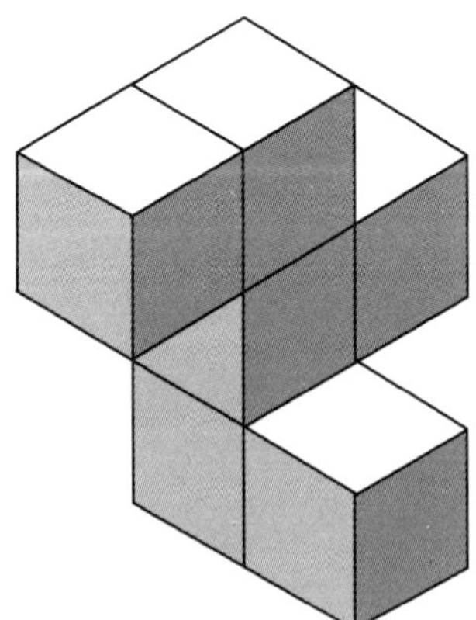

9.2 The 3-D View from Three Views of an Object

When we draw a 3-D view based on three views of an object on plain paper, we need a 3-dimensional imagination. We can first think about the front view, and then suppose the top view is "a hat" covering "the head" that is the front view. Finally, use the side view to fine-tune the image in your head. Sometimes this method might need to be repeated several times until the 3-D view matches the object's three views.

Example 1

Draw a 3-D view based on three views of an object.

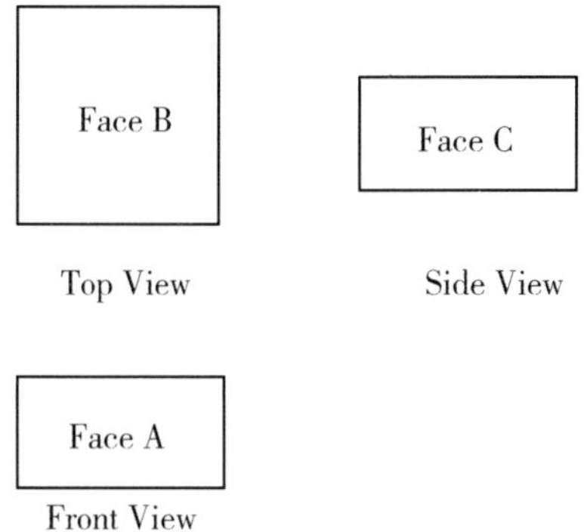

▶ *Solution*:

We can find four solutions *a*, *b*, *c* and *d*. All of these solutions are correct. The difference between the four 3-D views is the perspective from which we see the object. We need to understand how the three views match the surface of the 3-D view. Face D is the opposite side of Face B, and Face E is the opposite side of Face C.

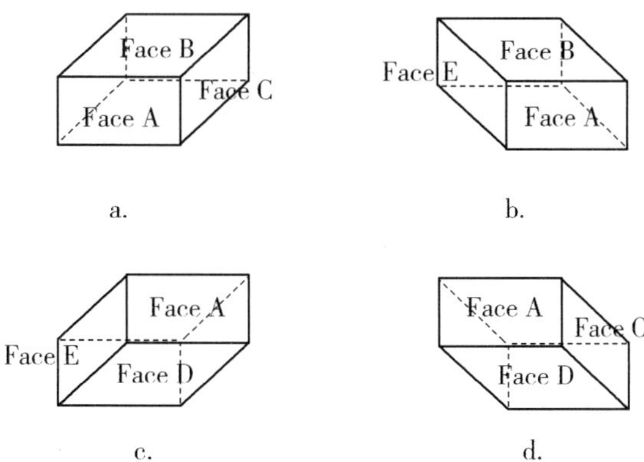

Example 2

Draw a 3-D view based on the three views of the object.

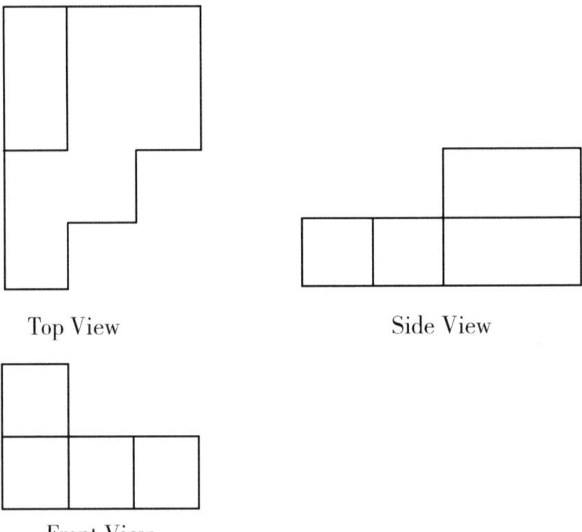

▶ *Solution*:

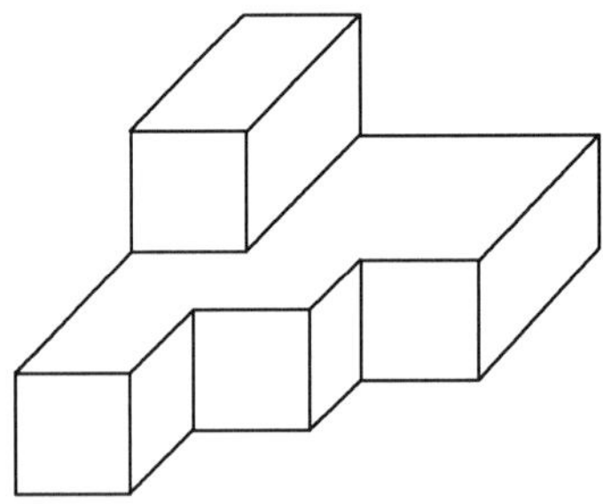

Practice

■Level A

1. Find the 3-D view which matches the three views.

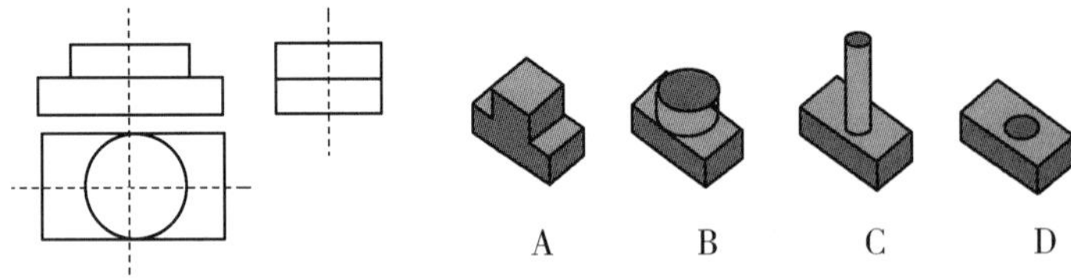

2. Draw a 3-D view based on the three views of the object.

Top View

Side View

Front View

3. Draw a 3-D view based on the three views of the object.

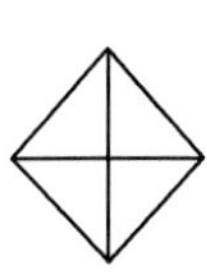
Top View

Side View

Front View

■Level B

1. Draw a 3-D view based on the three views of the object.

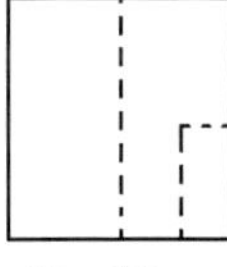
Top View

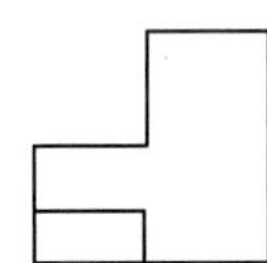
Side View

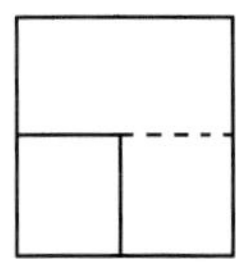
Front View

2. The object below was constructed from several cubes. Find how many cubes were used to make the object and draw the 3-D view.

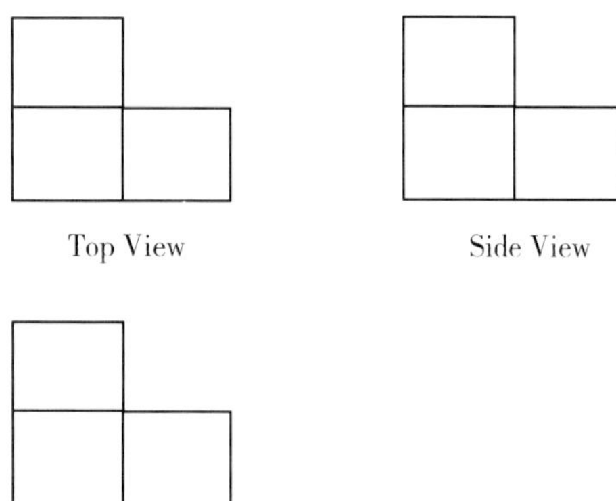

Top View Side View Front View

3. Draw a 3-D view based on the the three views of the object.

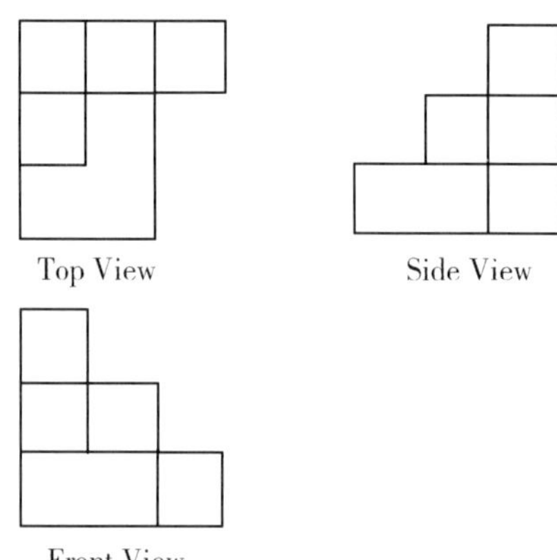

Top View Side View Front View

4. Draw a 3-D view based on the three views of the object.

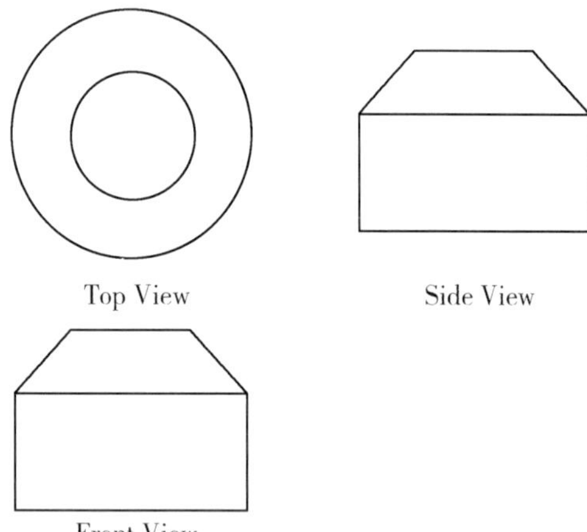

Top View Side View Front View

Level C

1. Draw a 3-D view based on the three views of the object.

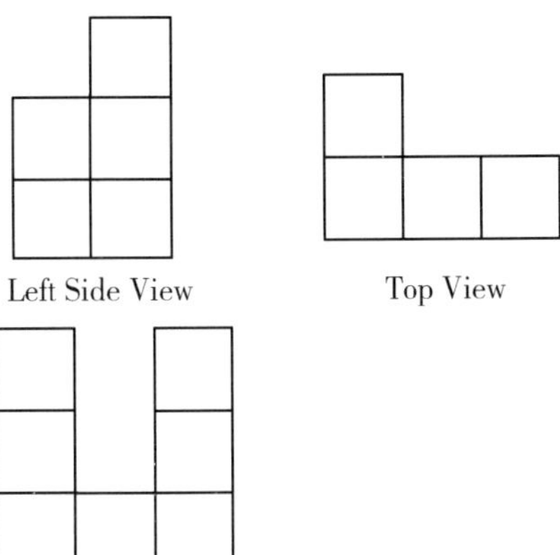

Left Side View

Top View

Front View

2. Draw a 3-D view based on the three views of the object.

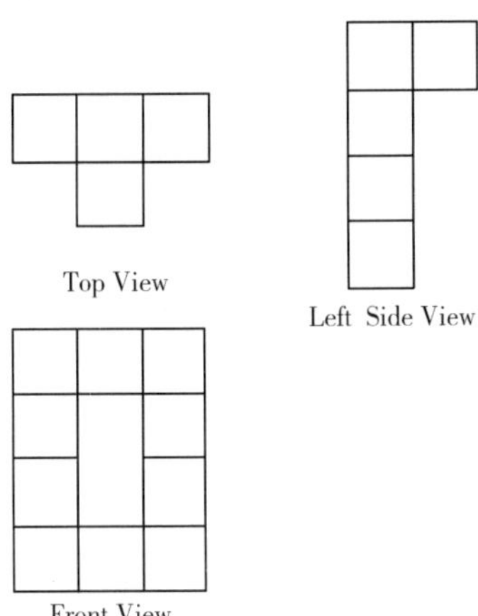

Top View

Left Side View

Front View

9.3 3-D Views and Three Views of Combined Objects

Example 1

Draw the 3-D view based on the following three views.

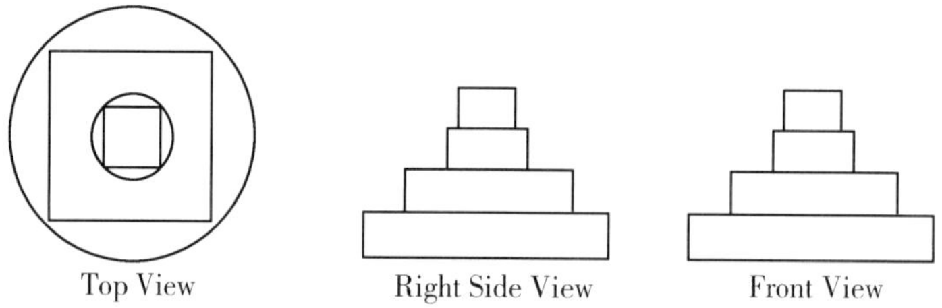

▶ *Solution*:

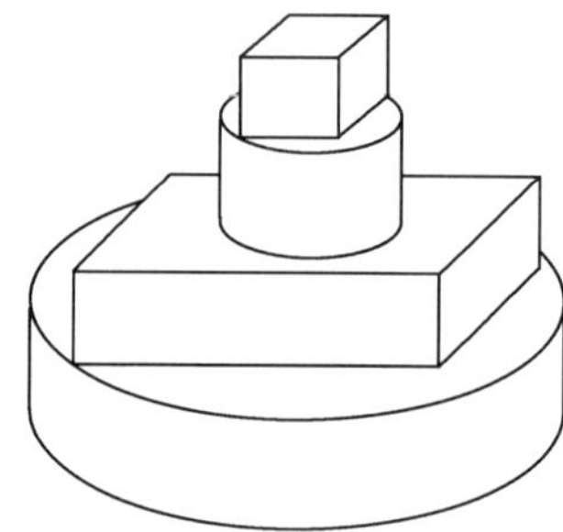

Example 2

Draw the top view, front view and side view based on the following 3-D view.

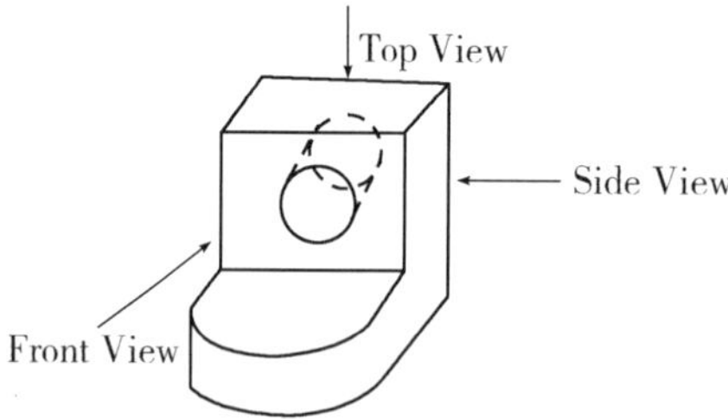

▶ *Solution*:

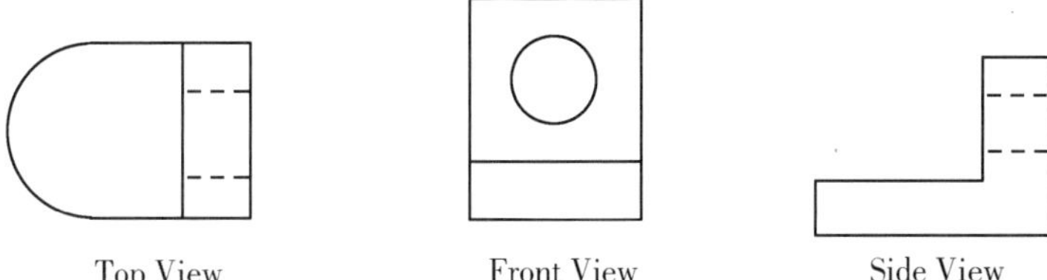

Practice

■Level A

1. Draw the top, front, and right side views based on the 3-D view.

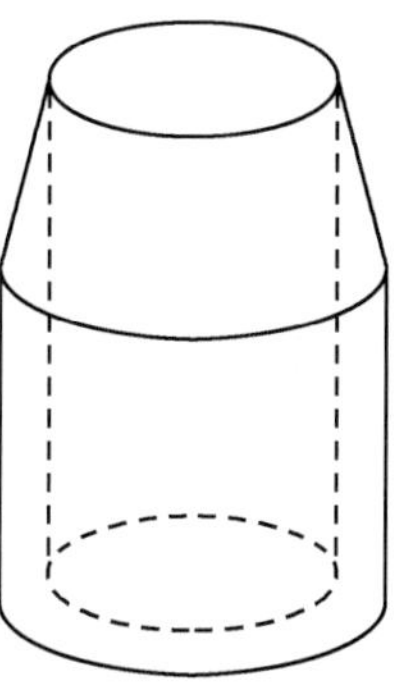

2. Draw the 3-D view based on the following three views.

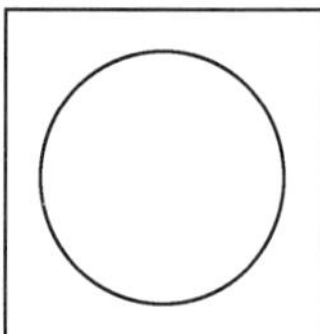

Top View

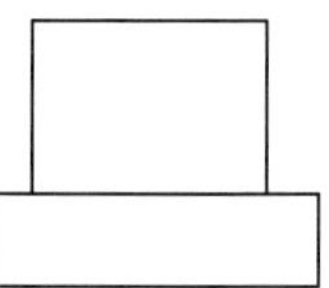

Front View

■Level B

1. Draw the top, front, and right side views based on the 3-D view.

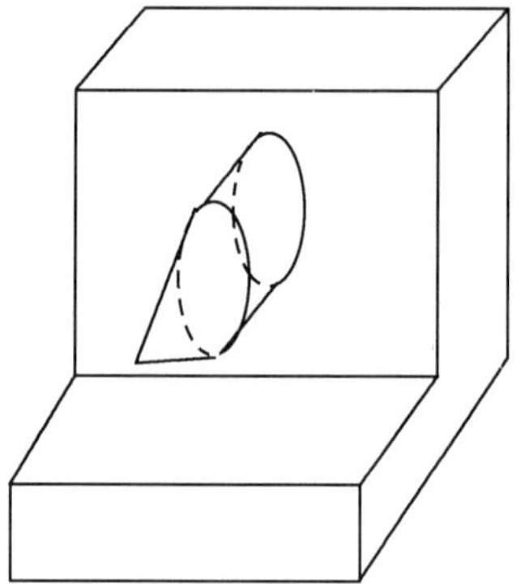

2. Draw the 3-D view based on the following three views.

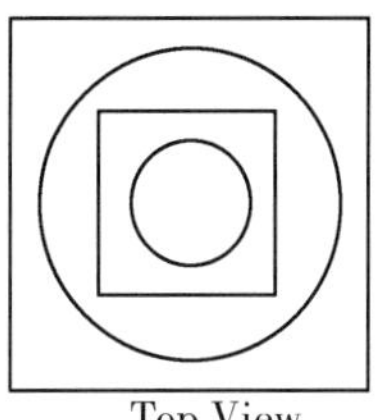
Top View

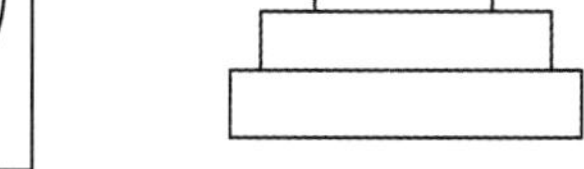
Right Side View

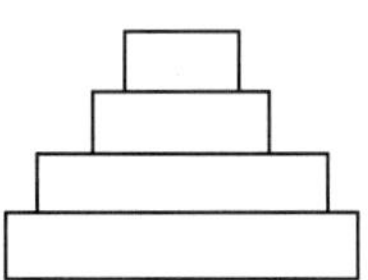
Front View

3. Draw the 3-D view based on the following three views.

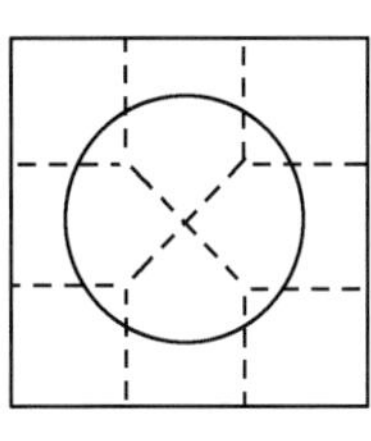
Top View

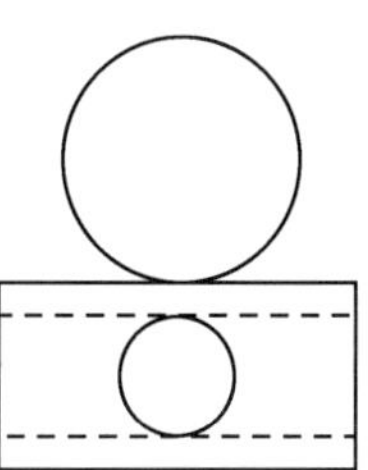
Right Side View

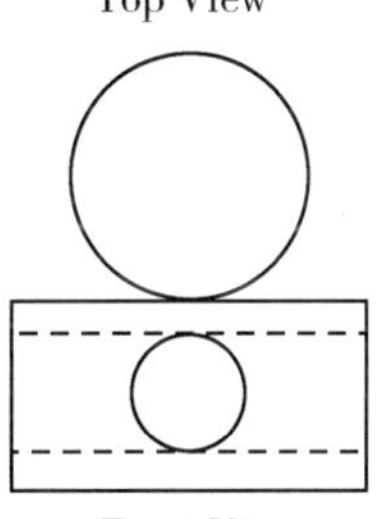
Front View

Chapter 9 TEST

1. Draw the 3-D view based on the following three views.

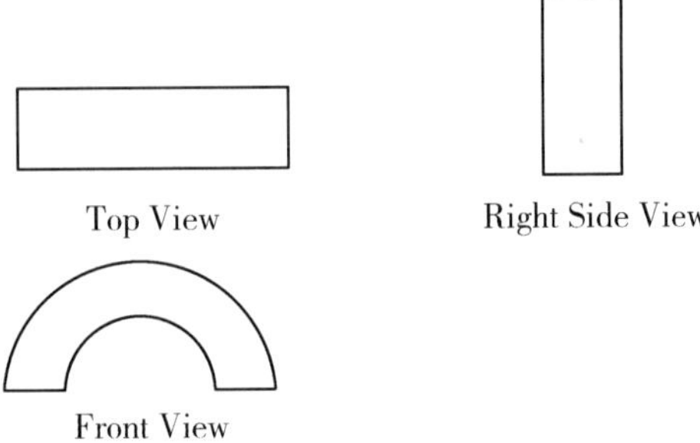

2. Draw the 3-D view based on the following three views.

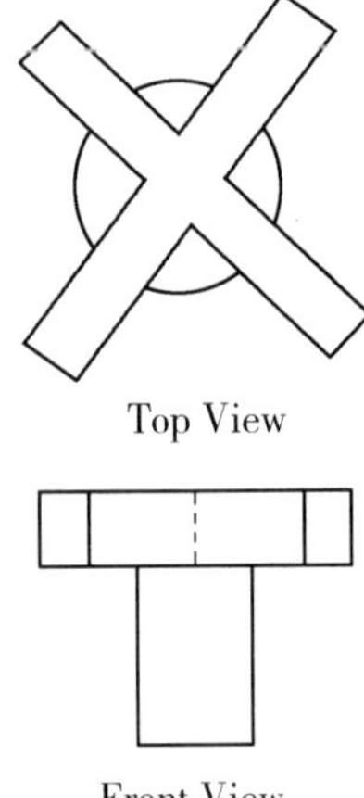

3. Draw the 3-D view based on the following three views.

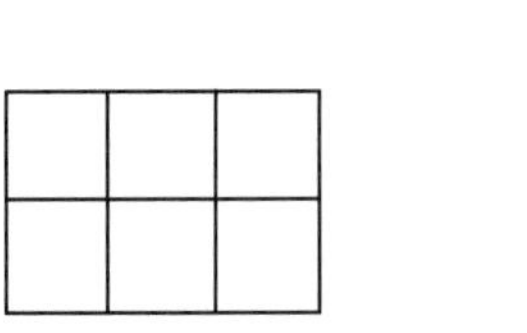

Top View

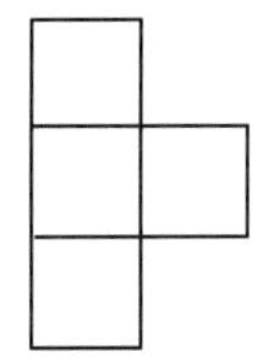

Left Side View

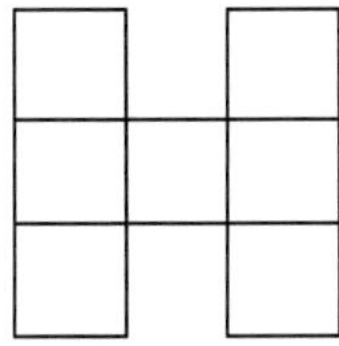

Front View

4. Draw the top, front and right side views based on the 3-D view.

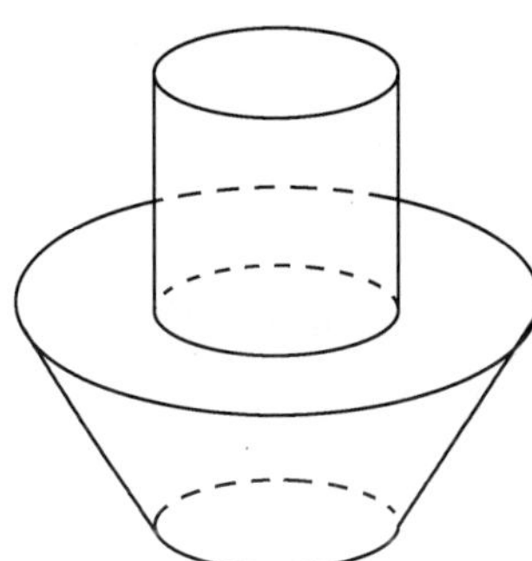

5. A die is a 6-sided cube with dots on its faces marking the numbers 1—6. Draw two possible nets of a die. Note that any two opposite faces of a die must be marked with numbers which add up to seven.

6. The diagram below is the front view of an object. The number which is indicated on each cube represents the number of cubes behind it. Draw a possible top view and right side view of the object.

2	3
	1

7. A corner is cut off of a rectangular prism. Draw a top view diagram of the prism after the corner was cut.

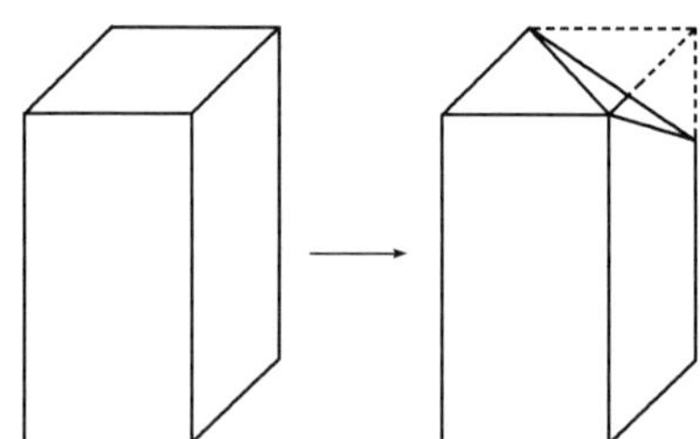

8. An object is shown below. A few different views of the object are shown as a, b, c, d. In each view, the number on each cube indicates the number of cubes behind it. Which of view does not match the object?

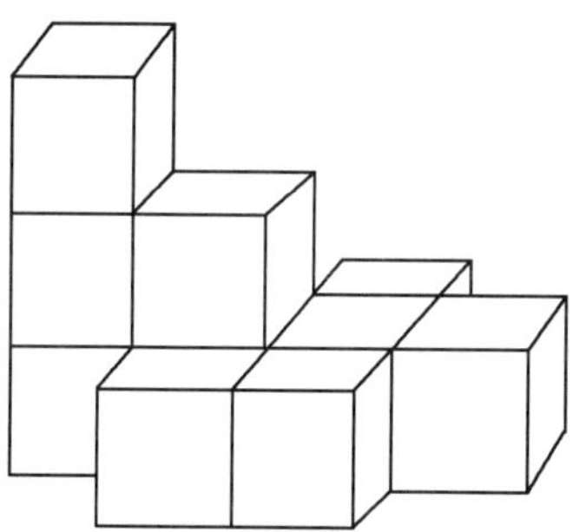

a)

	1	
	2	
1	4	2

b)

	1	
	2	
2	4	2

c)

1			
1	1		
1	2	3	1

d)

		1	
3	2	1	1
	1	1	

CHAPTER 10

Probability

10. 1 Probability

10. 2 Independent Events and
Dependent Events

Chapter 10 TEST

Chapter 10 Probability

10.1 Probability

Probability

Probability is a measure or estimation of how likely it is that an event will happen or how likely a statement is true. Probabilities are given by a value of 0 or 1. If the probability equals 0, then that means the event will not nearly happen. On the other hand, if the probability equals 1, then that means the event will always happen.

Example 1

Find the probability of getting a "7" while rolling a die numbered from "1" to "6".

▶ *Solution*:

P(7)=0, because the event is impossible. (P(7) denotes the probability of getting a "7").

Example 2

Find the probability of getting a number less than "7" while rolling a die numbered from "1" to "6".

▶ *Solution*:

P(1 to 6)=1, because the event will always happen.

Experimental Probability

Experimental Probability is the likelihood of an event estimated by repeating an experiment many times and observing the number of times the event happens. That number is divided by the total number of trials. The more the experiment is repeated, the more accurate the estimate is likely to be.

$$\text{Experimental Probability}=\frac{\text{number of times an event occurs}}{\text{total number of trials}}$$

Example 3

Bob tossed a coin 100 times and got heads 47 times. Find the experimental probability of getting heads.

▶ *Solution*:

Probability can be expressed as a fraction, decimal, or percent. The answer is $\frac{47}{100}$ or 0.47 or 47%.

Theoretical Probability

The number of possible ways an event can happen compared with all possible outcomes is called **Theoretical Probability.**

Example 4

Find the probability of getting heads while tossing a coin.

▶ *Solution*: $\frac{1 \text{ outcome with heads}}{2 \text{ possible outcomes}}=\frac{1}{2}$ or 0.5 or 50%

Example 5

Find the probability of getting an even number while rolling a die.

▶ *Solution*: $\frac{3}{6}$ or $\frac{1}{2}$.

Theoretical probability is written as a ratio of the number of favourable outcomes to the number of possible outcomes, assuming that all the outcomes are equally likely. In this chapter, we will focus on theoretical probability.

Example 6

A standard deck of 52 cards has four suits: hearts, diamonds, clubs and spades. Each suit has 13 cards composed of: ace, 2, 3, 4, 5, 6, 7, 8, 9, 10, jack, queen, and king. Hearts and diamonds are red; clubs and spades are black. If one card is randomly selected, find the probability of:

a) P(a red card) **b**) P(a face card)

c) P(a face club) **d**) P(a black ace)

▶ *Solution*:

a) There are 26 red cards in total, so P(a red card) $=\frac{26}{52}=\frac{1}{2}$.

b) Each suit has 3 face cards, which means there is a total of 12 face cards. So P(a face card) $=\frac{12}{52}=\frac{3}{13}$.

c) There are 3 face clubs. So P(a face club) $=\frac{3}{52}$.

d) There are two black aces: an ace of spades and an ace of clubs, so P(a black ace) $=\frac{2}{52}=\frac{1}{26}$.

Sample Space

The Sample Space is the collection of all possible results or outcomes of an event. A sample space can be written as a list of outcomes, a table, or a diagram.

Tree Diagram

A Tree Diagram is a tool to visually explain the events. It is a visual representation of "the sample space" for the event.

Example 7

A coin is thrown twice. Show the sample space by a tree diagram.

First toss	Second toss	Outcomes and Probability
H ($P=\frac{1}{2}$)	H	HH $P(HH)=\frac{1}{4}$
	T	HT $P(HT)=\frac{1}{4}$
T ($P=\frac{1}{2}$)	H	TH $P(TH)=\frac{1}{4}$
	T	TT $P(TT)=\frac{1}{4}$

Practice

■Level A

1. A fair die was rolled. Find the following probabilities.

a) P(an even number) **b)** P(less than 5)

c) P(prime number) **d)** P(less than 7)

e) P(an odd or prime number) **f)** P(greater than 2 and less than 5)

2. There were 750 tickets sold for a school raffle with only one grand prize. Jerry bought 5 tickets. Express the probability that Jerry win the grand prize as a fraction, a decimal and a percentage.

3. A bag contains 8 white balls, 13 black balls and 15 yellow balls. If you randomly take a ball from the bag, what is the probability of picking up the following?

a) a yellow ball

b) a black ball

c) a white ball

d) a green ball

e) a ball that is not black

f) a black or yellow ball

g) a white or yellow ball

h) a white or black ball

4. A class contains 12 boys and 14 girls. The teacher calls on a student at random to go to the chalkboard. What is the probability that the first person called is **a)** a boy? **b)** a girl?

5. A letter is chosen at random from a given word. Find the probability that the letter is a vowel (a, e, i, o, or u) if the word is:

a) APPLE

b) BANANA

c) MATHEMATICS

d) EDUCATION

e) PHOTOGRAPHY

f) PUBLICATION

■Level B

1. If you were to roll a 6-sided die twice,

a) sketch a table representing the sample space.

b) sketch a tree diagram representing the sample space.

c) determine the probability of getting a "1" first and a "2" second.

d) determine the probability of getting an odd number first and an even number second.

e) determine the probability of getting a prime number first and an odd number second.

f) determine the probability of getting an even number on both rolls.

g) determine the probability of getting the number "5" on both rolls.

h) determine the probability of getting a number greater than "4" on at least one roll.

i) determine the probability of getting a number less than "3" on both rolls.

2. A standard deck of 52 cards has four suits: hearts, diamonds, clubs and spades. Each suit has 13 cards composed of: ace, 2, 3, 4, 5, 6, 7, 8, 9, 10, jack, queen, and king. Hearts and diamonds are red; clubs and spades are black. If one card is randomly selected, find:

a) P(a black card)

b) P(a red face card)

c) P(a diamond or heart card)

d) P(not a face card)

e) P(a black card with a prime number)

f) P(a face card or a heart)

■Level C

1. Ken is practising putting. The probability of him holing the first putt is 0.4. Each time he holes a putt, the probability of him holing the next putt is increased by 25%. Each time he misses a putt, the probability of him missing the next putt is increased by 20%. Determine the probability that:

a) He holes the first two putts.

b) He misses the first two putts.

c) He holes two of the first three putts.

2. The probability of being caught in a traffic jam in the morning rush hour is 0.6 on any particular weekday. If the occurrence of traffic jams on different days is assumed to be independent of each other, determine the probability that there will be a traffic jam on at least two out of three weekdays.

10.2 Independent Events and Dependent Events

Independent Events

Two or more events are independent if the occurrence or non-occurrence of one does not affect the probability of the others occuring.

Example 1

Are these events independent?

Event A: getting a "5" when rolling a die.

Event B: getting tails when tossing a coin.

▶ *Solution*:

Event A and Event B are independent because getting tails while tossing a coin is not affected by getting a "5" while rolling a die.

P(A) denotes the probability of Event A occurring, and P(B) denotes the probability of event B occurring. If Event A and Event B are independent, then we have:

P(A and B)=P(A)×P(B)

Example 2

If we throw a die and flip a coin at the same time, find P (H and 3).

▶ *Solution*:

The sample space of rolling a die is 6, and the sample space of a coin is 2.

$P(H)=\frac{1}{2}$ and $P(3)=\frac{1}{6}$.

Because throwing a die and flipping a coin are independent events,

So P(A and B)=P(A)×P(B), $P\ (H \text{ and } 3)=P(H)\times P(3)=\frac{1}{2}\times\frac{1}{6}=\frac{1}{12}$.

Dependent Events

Two or more events are dependent if the probability of the occurrence of one event is affected by the occurrence of another event.

Example 3

Tommy picked two cards from a standard deck of 52 cards, without putting the first card back.

Event A: the first card is a "7".

Event B: the second card is a "7".

Are these two events independent or dependent?

▶ *Solution*:

These two events are dependent. If Event A occurs (the first card is a "7") then there are only three "7" cards left. If Event A doesn't occur (the first card is not a "7") then there will still be four "7" cards left.

Therefore, the probability of Event B occuring will be affected by Event A.

Venn Diagrams

A Venn diagram is another great way to represent the relationship between two or more events.

Two events A and B are **mutually exclusive** if they cannot occur at the same time. When drawing the Venn diagram for actually exclusive events, we draw two regions that do not overlap.

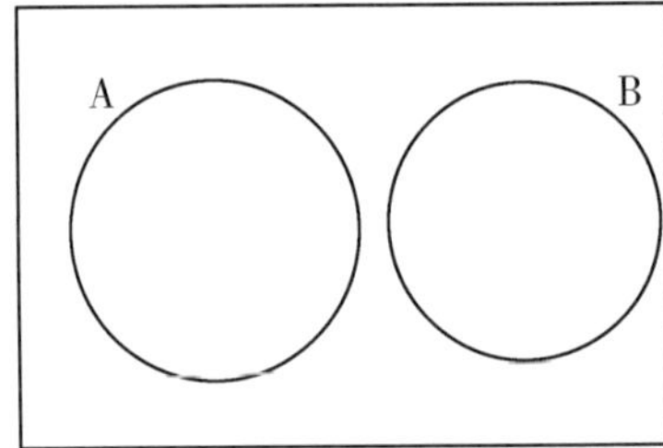

Example 4

Bobby picked one card from a deck of cards.

Event A: getting a red "2" from a standard deck of cards.

Event B: getting a red face card from a standard deck of cards.

Are these events mutually exclusive?

▶ *Solution*:

Event A and Event B are mutually exclusive because a red card cannot be a "2" and a face card at the same time.

Two events A and B are not mutually exclusive when they can occur at the same time. When drawing the Venn diagram for events that are not mutually exclusive, we draw two regions that have an overlap.

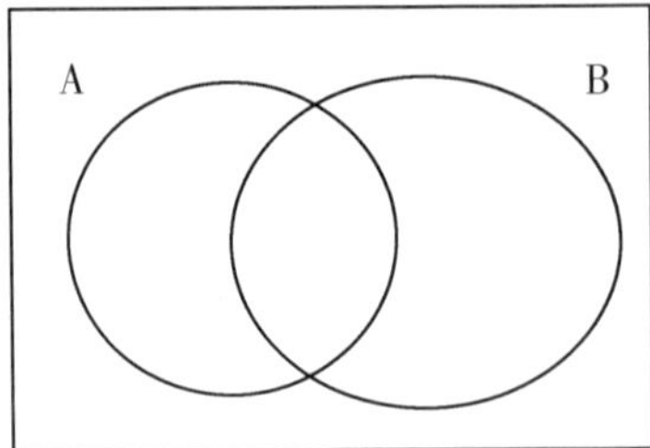

Example 5

A student is chosen from your class.
Event A: the student is left-handed.
Event B: the student is a female.
Are these events mutually exclusive?

▶ *Solution*:

In this case, it is possible to find a female student who is left-handed in your class. Event A and Event B are not mutually exclusive.

Complementary Events

The complement of the likelihood of an event is the likelihood that the event will not happen. For example, if Event A is drawing a five from a standard deck of cards, then the complement of Event A is not drawing a five. In this case, the probability of Event A is $\frac{4}{52}$, so the probability of the complement of Event A is $1-\frac{4}{52}=\frac{48}{52}$.

Practice

■Level A

1. A coin is tossed 3 times. What is the probability of getting the following?

a) three heads in a row

b) three tails in a row

c) one head

d) two heads

e) at least one tail

f) at least two heads

2. A die is tossed twice. What is the probability of getting the following?

a) greater than "4" on both rolls

b) the number "2" on both rolls

c) a sum greater than 8

d) a sum less than 7

e) the same number on both rolls

f) a sum that is a prime number

g) an even number on both rolls

h) a sum greater than 12

■Level B

1. A bag contains 9 balls: 2 red balls, 4 yellow balls and 3 blue balls. A ball is picked randomly and put back, and then another ball is picked. Find the probability of getting:

a) both blue

b) both yellow

c) one red, one yellow

d) first red, second yellow

e) not blue

f) not red

g) not red or blue

h) at least one red

i) at least one blue

j) first yellow, second blue

2. There is a pattern in the following diagrams. If a ball is randomly taken out from diagram n, what is the probability of taking out a black ball?

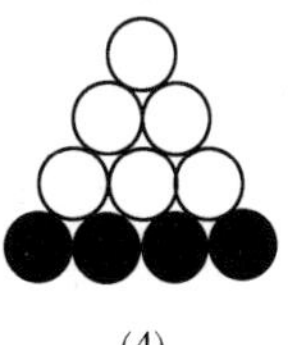

(1) (2) (3) (4) …

3. A bag contains 9 balls: 2 red balls, 4 yellow balls and 3 blue balls. One ball is picked randomly and then another ball is picked without putting the first one back. Find the probability of getting:

a) both blue

b) both yellow

c) one red, one yellow

d) first red, second yellow

e) not blue

f) not red

g) not red or blue

h) at least one red

i) at least one blue

j) first yellow, second blue

■Level C

1. 20% of the students in Mr. Johnson's Pre-calculus 12 class earned over 85%; 52% of his students earned over 72%; and 12% earned less than 50%. What is the probability that a randomly selected student from Mr. Johnson's Pre-calculus 12 class earned 50% or above?

2. A survey reveals that 81% of teenagers in British Columbia listen to pop and 14% listen to classical music. 13% listen to neither type of music. What is the probability that a randomly selected British Columbia teenager would listen to both types of music?

Chapter 10 TEST

1. A bag contains 8 white balls, 13 black balls and 15 yellow balls. If you randomly take a ball from the bag, put it back and then take another ball, find the probability of picking up:

a) one yellow, one black

b) first one yellow, second one black

c) two white ones

d) two black ones

e) no black ones

f) one black, one yellow

g) both yellow

h) one white, one yellow

2. A bag contains 8 white balls, 13 black balls and 15 yellow balls. If you randomly take one ball from the bag and then take another ball without putting back the first one, what is the probability of picking up the following?

a) one yellow, one black

b) first one yellow, second one black

c) two white ones

d) two black ones

e) no black ones

f) one black, one yellow

g) both yellow

h) one white, one yellow

3. A die is tossed twice. What is the probability of getting the following?

a) less than "3" on both rolls

b) the number "5" on both rolls

c) an odd number on both rolls

d) a sum greater than 1

e) a product that is even

f) a sum that is even

g) a sum that is a prime number

h) a sum that is a multiple of 3

i) a difference of less than 4

j) a difference that is a prime number

4. A standard deck of 52 cards has four suits: hearts, diamonds, clubs and spades. Each suit has 13 cards composed of: ace, 2, 3, 4, 5, 6, 7, 8, 9, 10, jack, queen, and king. Hearts and diamonds are red; clubs and spades are black. If one card is randomly selected, find:

a) P(a club card)

b) P(a black face card)

c) P(a diamond or spade card)

d) P(NOT a face card)

e) P(a red card with a prime number)

f) P(a black card with an even number)

5. In a standard deck of 52 cards, if one card is randomly selected, put back, and then another card is selected, find:

a) P(one red, one black)

b) P(first red, second black)

c) P(both red)

d) P(no face card)

e)P(both face cards)

f)P(first club，second heart)

g)P(both cards are kings)

h)P(both are number cards)

6. In a standard deck of 52 cards，if one card is randomly selected and then another card is selected without putting the first card back，find the probability of：

a)P(one red，one black)

b)P(first red，second black)

c)P(both red)

d)P(no face card)

e)P(both face cards)

f)P(first club，second heart)

g) P(both cards are kings)

h) P(both greater than 5, excluding jack, queen, and king)

7. There are two bus routes between Vincent's home and downtown where he works out. Vincent uses Route A two-thirds of the time, and Route B one-third of the time. On Route A, 10% of the buses are late, and on Route B, 20% of the buses are late. What is the probability of Vincent being late for his work out on any particular day?

8. The probability that a new car will not require service before a year has passed is 0.85, and the probability it will not require service in 2 years is 0.10. At the end of the first year, a car did not need servicing. What is the probability that the car will not require service in the second year?

APPENDIX

MATHEMATICS 8 WORKBOOK

ANSWERS TO PRACTICE AND CHAPTER TESTS

ANSWERS CHAPTER 1

1.1 Answer Key

Level A

1.

	Power	Base	Exponent	Standard form
a)	3^2	3	2	3×3
b)	2^4	2	4	$2\times2\times2\times2$
c)	4^3	4	3	$4\times4\times4$
d)	5^2	5	2	5×5
e)	8^3	8	3	$8\times8\times8$
f)	9^2	9	2	9×9

2. a) 5^2 b) 9^2
   c) 10^2 d) 13^2
   e) 121^2 f) 32^2
3. a) $102^2=10\ 404$ b) $0.5^2=0.25$
   c) $1.2^2=1.44$
4. a) 3^2 b) 3^4 c) 3^3
5. a) 2^5 b) 3^4 c) 3^8
6. 16 m
7. 500 cm^2
8. a) 2.3×10^4 b) 3.0×10^9
   c) 1.0×10^6 d) 1.0×10^8
   e) 3.2×10^6 f) 5.7×10^5
9. a) 1000 b) 2000
   c) 50 000 d) 33.3
   e) 2.5 f) 52 000
10. a) 2.0×10^{-2} b) 5.0×10^{-2}
    c) 3.0×10^{-4} d) 5.786×10^4
    e) 6.02×10^{-5} f) 5.003×10^{-4}
11. a) 2100 b) 0.0312
    c) 0.000012 d) 0.000017285
    e) 0.000002519 f) 0.000005101

Level B

1. a) $(\frac{4}{5})^3=0.512$ b) $0.7^3=0.343$
   c) $13^3=2197$
2. a) 3^6 b) 2^{17}
   c) 9^6
3. a) $2^4, 2^5, 2^6$ (16, 32, 64)
   b) 14, 17, 20
   c) $5^4, 5^5, 5^6$ (625, 3125, 15 625)
   d) $2\times3^4, 2\times3^5, 2\times3^6$ (162, 486, 1458)
4. a) 7×3^2 b) 3×7^2
   c) $7^2\times3^2$
5. a) 5.2×10^2 b) 9.7×10^3
   c) -1.8×10^3 d) 4.8×10^4
6. a) 6.35×10^{-2} b) 6.13×10^{-3}
   c) 7.69×10^{-3} d) 1.99×10^{-4}

Level C

1. a) -125 b) 81
   c) 64
2. a) 1 b) 0.1
   c) -0.25
3. a) 7×5^2 b) $9^2\times5^2$
   c) 7^4
4. a) 1.8312×10^3 b) 3.2512×10^{-7}
   c) 2.2272×10^1 d) 2.41724×10^3
5. a) $-1.296\times10^2=-129.6$
   b) $8.349\times10^0=8.349$
   c) $2.4192\times10^4=24\ 192$
   d) $5.0\times10^5=500\ 000$

1.2 Answer Key

Level A

1. a) 60% b) 50%
   c) 100%
2. a) 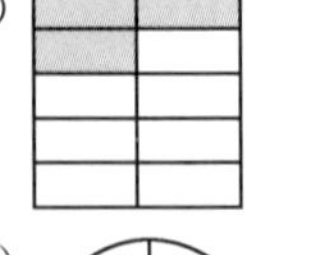 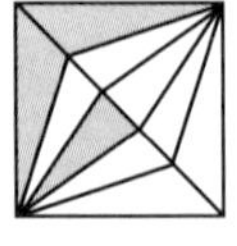
   b) 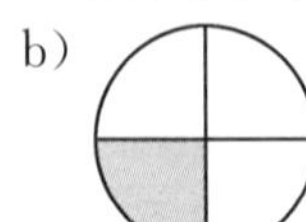 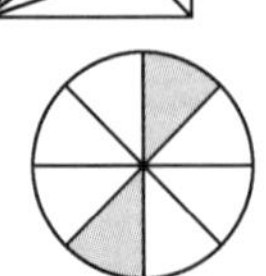

3. a) 50% b) 70%
   c) 90% d) 98%
   e) 25% f) 50%
4. a) $\frac{1}{4}$ b) $\frac{1}{7}$
   c) $\frac{1}{11}$ d) 1 : 9
   e) 1 : 15 f) 1 : 5
5. a) 18 b) 21
   c) 8 d) 225
   e) $\frac{49}{9}$ f) 12
6. Jack can run $4\frac{2}{3}$ km (4.667 km) in 70 minutes. Jack takes 52.5 minutes to run 3.5 km.

Level B

1. a) 50% b) 25%
   c) 20% d) 60%
   e) 75% f) 40%
2. a) 0.23 b) 0.34
   c) 0.56 d) 0.08
   e) 0.05 f) 0.09
3. 33%
4. a) 57.4 b) 22.4
   c) 7.3 d) 5.3
   e) 8.3 f) 11.7
5. a) 6 b) 20
   c) 60 d) $\frac{20}{9}$
6. $38.57

Level C

1. a) 2330.00% b) 177.78%
   c) 255.56% d) 533.33%
   e) 202.02% f) 301.00%
   g) 0.92% h) 92.93%
   i) 92.93% j) 27.84%
   k) 86.67% l) 109.09%
2. x: previous pay
   $x(1+25\%)=1000$
   $x=\frac{1000}{1+25\%}=\frac{1000}{1.25}=\$800$
3. a) $\frac{\$2.86}{4\text{ kg}}=\$0.715/\text{kg}$
   $\frac{\$7.81}{11\text{ kg}}=\$0.71/\text{kg}$ (better deal)
   b) $\frac{18\text{ h}}{13\text{ d}}=1.38$ h/day (better deal assuming students like do less homework each day)
   $\frac{21\text{ h}}{15\text{ d}}=1.4$ h/day
4. a) $\frac{x}{5}=\frac{17}{35}, x=\frac{5\times17}{35}=\frac{17}{7}$
   $\frac{3}{5}=\frac{y}{35}, y=\frac{3\times35}{5}=21$
   b) $\frac{x}{\frac{3}{5}}=\frac{0.75}{3.5}, x=\frac{\frac{3}{5}\times0.75}{3.5}=\frac{9}{70}$
   $\frac{1}{2}:\frac{3}{5}=y:3.5, \frac{3}{5}y=\frac{1}{2}\times3.5, y=\frac{35}{12}$
   c) $19:x=38:55, 38x=19\times55, x=\frac{55}{2}$
   $19:77=38:y, 19y=77\times38, y=154$
   d) $\frac{2}{3}:0.8=x:0.2, 0.8x=\frac{2}{3}\times0.2, x=\frac{1}{6}$
   $0.8:y=0.2:16, 0.2y=0.8\times16, y=64$
5. x: the distance sound travels in 4.5 seconds in water
   $\frac{1460\text{ m}}{1\text{ s}}=\frac{x}{4.5\text{ s}}$
   $x=1460\times4.5=6570$ (m)
   t: the time sound takes to cross 438 km in the sea
   $\frac{1460\text{ m}}{1\text{ s}}=\frac{438\,000\text{ m}}{t}$
   $t=\frac{438\,000}{1460}=300$ s

1.3 Answer Key

Level A

1. 1, 4, 9, 16, 25, 36, 49, 64, 81, 100, 121, 144, 169, 196, 225, 256, 289, 324, 361,

400

2. a) 6^2 b) 7^2
   c) 8^2 d) 9^2
   e) 12^2 f) 15^2
   g) 11^2 h) 16^2
   i) 17^2 j) 14^2
   k) 18^2 l) 19^2
3. a) 0.1^2 b) 0.03^2
   c) 0.012^2 d) 1.4^2
   e) 0.09^2 f) 0.17^2
   g) $(\frac{6}{7})^2$ h) $(\frac{11}{15})^2$
   i) $(\frac{12}{18})^2$ j) $(\frac{16}{14})^2$
   k) $(\frac{15}{20})^2$ l) $(\frac{14}{19})^2$
4. a) 8 b) 12
   c) 19 d) 200
   e) 130 f) 160
   g) 0.6 h) 1.5
   i) 1.7 j) $\frac{1}{2}$
   k) $\frac{13}{50}$ l) $\frac{17}{18}$
5. $l=5$ cm, $w=5$ cm, $h=7$ cm

Level B

1. a) $2^3\times3^2$ b) $2^2\times13$
   c) $2^4\times5$ d) 3^5
   e) $2^2\times3^3$ f) $2^2\times5^2$
   g) 11^3 h) $3^2\times2\times5$
   i) $3^2\times2\times37$ j) $2^4\times3^3$
   k) $2^3\times11$ l) 11×13
2. a) 2.2, 2.236 b) 4.9, 4.8990
   c) 7.1, 7.1414 d) 8.4, 8.3666
   e) 11.5, 11.489 f) 12.5, 12.4900
   g) 0.2, 0.2449 h) 10.0, 9.9499
   i) 7.3, 7.3485 j) 10.5, 10.4881
   k) 8.2, 8.1854 l) 3.3, 3.3166
3. Bob is 18, and Tim is 9.

Level C

1. a) -125 b) 256
   c) -625 d) $\frac{1}{25}$
   e) -1000 f) $-\frac{1}{100}$
   g) 0.09 h) $\frac{100}{9}$
   i) -0.09 j) -9
   k) $-\frac{1}{9}$ l) $-\frac{1}{9}$
2. a) n/a b) n/a
   c) 4 d) n/a
   e) n/a f) $\frac{1}{13}$
   g) n/a h) n/a
   i) 2 j) n/a
   k) 0.25 l) n/a
3. a) 1.73 b) 2.24
   c) 1.41 d) 3.61
   e) 2.65 f) 2.45
   g) 5.20 h) 2.83
   i) 3.16

1.4 Answer Key

Level A

1.

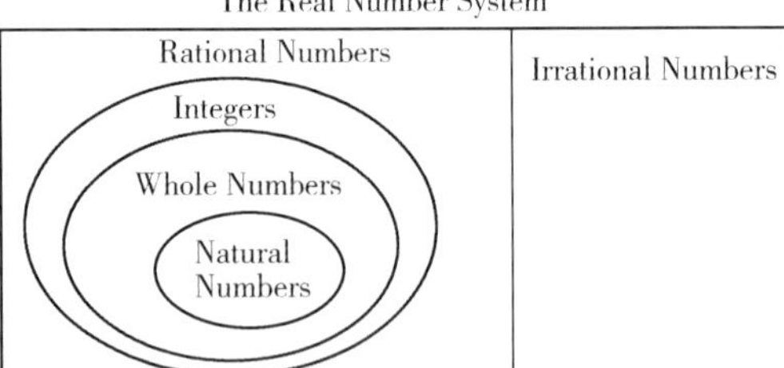

2.

		Set of Numbers					
	Number	N	W	I	Ra	Ir	R
a)	-4			√	√		√
b)	0		√	√	√		√
c)	9	√	√	√	√		√

d)	$0.\overline{37}$				√		√
e)	0.37				√		√
f)	$\sqrt{3}$					√	√
g)	$4\sqrt{3}$					√	√
h)	π					√	√
i)	$\sqrt{\frac{9}{16}}$				√		√
j)	-0.98				√		√

Level B

1. a) $\frac{17}{100}$ b) $-\frac{1005}{1000}$
 c) $\frac{25}{1000}$ d) $\frac{2}{3}$
 e) $\frac{3}{5}$ f) $\frac{15}{3}$
 g) $\frac{47}{100}$ h) $\frac{3}{100}$
 i) $\frac{9}{10}$ j) $\frac{107}{100}$
 k) $\frac{213}{100}$ l) $\frac{3172}{1000}$
2. a) $0.5<0.\overline{5}$ b) $0.9<0.\overline{9}$
 c) $1>0.\overline{9}$ d) $-0.3<0.\overline{3}$
 e) $-0.3^0<(-3)^0$ f) $\frac{1}{3}=0.\overline{3}$
 g) $0.8^{-1}>0.\overline{8}$ h) $-0.8^{-1}<-0.8^1$
 i) $-0.\overline{08}<-\frac{4}{50}$ j) $-5^2<(-5)^2$
 k) $-5^2<(-5)^{-2}$ l) $-5^{-2}<(-2)^{-2}$

Level C

1. a) $\frac{135}{999}$ b) $\frac{27}{200}$
 c) $\frac{67}{495}$ d) $\frac{7}{6}$
 e) $\frac{29}{25}$ f) $\frac{1121}{999}$
 g) n/a h) $\frac{31\ 313\ 133}{10\ 000\ 000}$
 i) $\frac{469\ 697}{150\ 000}$ j) $\frac{1}{2}$
 k) n/a l) $-\frac{2}{3}$
2. a) 2.65 b) 1.41
 c) 3.74 d) 4.58
 e) 1.73 f) 2.24
 g) 10.05 h) 1.00
 i) 3.18 j) 15.20
 k) 1.52 l) 4.81

Chapter 1 TEST

1.

	Power	Base	Exponent	Standard form
a)	6^2	6	2	6×6
b)	$(-2)^4$	-2	4	$(-2)\times(-2)\times(-2)\times(-2)$
c)	1^8	1	8	$1\times1\times1\times1\times1\times1\times1\times1$
d)	$(-5)^2$	-5	2	$(-5)\times(-5)$
e)	0.4^3	0.4	3	$0.4\times0.4\times0.4$
f)	$(-11)^2$	-11	2	$(-11)\times(-11)$

2. a) 12^3 b) 11^2
 c) 7^4 d) 21^5
 e) 202^2 f) 15^3
3. a) $5.5^2=30.25$ b) $(\frac{2}{3})^2=\frac{4}{9}$
 c) $(\frac{3}{4})^2=\frac{9}{16}$
4. a) 3^1 b) 3^5
 c) 3^6
5. a) 2^5 b) 3^5
 c) 2^{301}
6. a) 5.6789×10^7 b) 1.001×10^8
 c) 1.2021×10^8
7. a) −25 600 b) −2 700 000
 c) 536 000 d) 22 780
 e) 31 850
8. a) 3.571×10^{-2} b) 8.571×10^{-2}
 c) $3.\overline{03}\times10^{-2}$ d) -9.87×10^{-4}
 e) 7.03×10^{-3} f) -1.0001×10^{-3}
9. a) −0.00251 b) −0.037

c) -0.004281 d) 0.303

e) 0.1001

10. a) $(\frac{1}{2})^4=\frac{1}{16}$

b) $(0.3)^4=0.0081$

c) $(\frac{3}{4})^3=\frac{27}{64}$

11. a) $3^8=6561$ b) $2^{16}=65\ 536$

c) 9

12. a) $2^2\times7^2$ b) 7^3

c) 3^4 d) 3^6

13. a) 1.125×10^8 b) 8.37×10^6

c) 4.4 d) 2

14. a) 7.69×10^{-3} b) 1.99×10^{-4}

15. a) -625 b) 64

c) -512

16. a) $10\times2\times2\times2=80$

b) $(-\frac{1}{2})\times(-\frac{1}{2})=\frac{1}{4}$

c) $-(\frac{1}{3})\times(\frac{1}{3})=-\frac{1}{9}$

17. a) $5^2\times7^2\times9$ b) $5^2\times7^2$

c) $5^2\times7\times9$

18. a) 4.675×10^{-2} b) 8.1×10^3

c) 8.7663×10^5 d) 1.2133×10^{-2}

19. a) -9.8×10^{-4} b) 4.0×10^2

c) -3.0×10^4 d) -7.848×10^{-3}

20. a) 55.56% b) 33.33%

c) 75%

21. a) b)

22. a) 115% b) 235%

c) 677% d) 5%

e) 1010% f) 809%

23. a) 6∶1 b) 1∶12

c) 1∶28 d) 5∶1

e) 1∶9 f) 20∶1

24. a) 120 b) 5

c) 10 d) 25

e) 0.6 f) 98

25. a) 97% b) 77%

c) 100% d) 65%

e) 84% f) 98%

26. a) 1.16 b) 2.20

c) 1.00 d) 0.005

e) 0.0025 f) 0.1001

27. a) 1.9 b) 2.6

c) 0.2 d) 2.4

e) 1.2

28. a) $x=\frac{10}{3}$ b) $x=288$

c) $x=36$ d) $x=363$

29. a) 12 books for \$20.00

b) \$625 for 5 nights

30. a) $x=29.4$

b) $y=24.0$

c) $x=15.0, y=12.5$

d) $x-20.0, y=2.2$

31.

		Set of Numbers					
	Number	N	W	I	Ra	Ir	R
a)	-18			√	√		√
b)	0		√	√	√		√
c)	$0.\overline{29}$				√		√
d)	$\sqrt{7}$					√	√
e)	π					√	√
f)	$\sqrt{\frac{4}{25}}$				√		√

32. a) $\frac{4}{11}$ b) $\frac{6}{13}$

c) $\frac{3}{500}$ d) $\frac{255}{999}=\frac{85}{333}$

e) $\frac{255}{1000}=\frac{51}{200}$ f) $\frac{23}{90}$

33. a) 3.61 b) 4.58

c) 6.78

ANSWERS CHAPTER 2

2.1 Answer Key

Level A

1. a)0.88 b)0.76
   c)0.35 d)0.69
   e)0.0006 f)0.0058
   g)0.012 h)0.0406
   i)2.07 j)3.14
   k)5.15 l)9.03

2. a)$\frac{78}{100}=\frac{39}{50}$ b)$\frac{65}{100}=\frac{13}{20}$
   c)$\frac{48}{100}=\frac{12}{25}$ d)$\frac{29}{100}$
   e)$\frac{55}{100}=\frac{11}{20}$ f)$\frac{75}{100}=\frac{3}{4}$
   g)$\frac{42}{100}=\frac{21}{50}$ h)$\frac{30}{100}=\frac{3}{10}$
   i)$\frac{355}{1000}=\frac{71}{200}$ j)$\frac{224}{1000}=\frac{28}{125}$
   k)$\frac{155}{1000}=\frac{31}{200}$ l)$\frac{655}{1000}=\frac{131}{200}$

3. a)64% b)57%
   c)96% d)32%
   e)5% f)25%
   g)15% h)8%
   i)1123% j)2050%
   k)1905% l)12 330%

Level B

1. a)37.5% b)75%
   c)6.25% d)31.25%
   e)48.65% f)86.67%
   g)44.44% h)42.86%
   i)32% j)1.85%
   k)31.4% l)47.37%

2. a)$2\frac{37}{1000}$ b)$\frac{1}{8}$
   c)$15\frac{85}{100}=15\frac{17}{20}$ d)$10\frac{5}{1000}=10\frac{1}{200}$
   e)$1\frac{4505}{10\ 000}=1\frac{901}{2000}$ f)$\frac{2368}{10\ 000}=\frac{148}{625}$
   g)$1\frac{607}{10\ 000}$ h)$6\frac{521}{1000}$
   i)$\frac{65}{100}=\frac{13}{20}$ j)$\frac{25}{100\ 000}=\frac{1}{4000}$
   k)$\frac{125}{100\ 000}=\frac{1}{800}$
   l)$\frac{10\ 015}{1\ 000\ 000}=\frac{2003}{200\ 000}$

3. 89%

4. $2\frac{1}{4}$ m or $\frac{9}{4}$ m

5. 66.67%

6. $\frac{1}{4}$

7. 0.75

8. 12.5%

9. $0.8<0.87=87\%<\frac{7}{8}$

10. 4 : 1

11. a)$\frac{7}{6}$ m b)1.17 m

12. 55.56%

13. 60%, $\frac{1}{2}$, 0.4

Level C

1. a)> b)> c)<

2. $5000\times 2.25\%=5000\times 0.0225=\$112.5$

3. 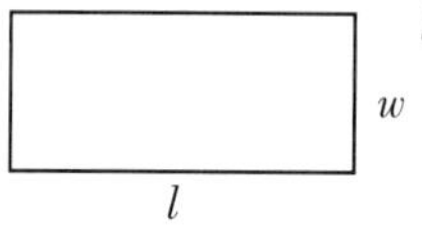

$l : w=5 : 2$

$\frac{l}{w}=\frac{5}{2}$

$l=\frac{5}{2}w$

$2w+2l=280$

$2w+2\times\frac{5}{2}w=7w=280$

$w=40$ m

$l=\frac{5}{2}\times 40=100$ m

area $= l \times w = 4000\ m^2$

$16 \times 4000 = 64\ 000$ kg

2.2 Answer Key

Level A

1. a) 84% b) 33.33%
   c) 140% d) 175%
   e) 7.5 f) 1.15
   g) 102 h) 105
2. a) 128.57 b) 100
   c) 65.31 d) 27.27
   e) $120 f) $72
   g) 33.75 kg h) 625 cm
3. B=350
4. 25%

Level B

1. a) 70.4% b) 23.94%
   c) 24.89% d) 12.12%
   e) 101.5 f) 0.855
   g) 5130 h) 26.66
2. 180
3. A $= 5333\frac{1}{3}$
4. 32
5. 20%
6. 1400 tons
7. 80%
8. Cup B
9. a) 0.666, 67%, $\frac{3}{4}$
   b) 3.75%, 37.5%, $\frac{4}{9}$
   c) 12.9%, 1.27, 128.59%
   d) $\frac{1}{4}$, 25.1%, 24.5

Level C

1. $\frac{200}{2500} = 0.08 = 8\%$
2. a) $(1-25\%)x = 72$
   $(1-0.25)x = 72$
   $0.75x = 72$
   $x = \frac{72}{0.75} = 96$

   b) $x - 40\%x = 5.04$
   $x(1-40\%) = x(1-0.4) = 0.6x = 5.04$
   $x = \frac{5.04}{0.6} = 8.4$

   c) $x \div (1-40\%) = 3.6$
   $x \div (1-0.4) = x \div 0.6 = 3.6$
   $\frac{x}{0.6} = 3.6$
   $x = 3.6 \times 0.6 = 2.16$

   d) $3x - 25\%x = 12$
   $(3-25\%)x = 12$
   $(3-0.25)x = 12$
   $2.75x = 12$
   $x = \frac{12}{2.75} = \frac{48}{11} = 4.36$
3. x: tons of wheat originally stored at Warehouse B
   $\frac{110-10}{x+10} = 40\%$
   $100 = 40\%(x+10)$
   $= 0.4(x+10)$
   $= 0.4x+4$
   $0.4x = 96$
   $x = \frac{96}{0.4} = 240$ tons

2.3 Answer Key

Level A

1. GST= $11.50; PST= $16.10;
   Total= $27.60
2. GST= $4.00; PST= $5.60;
   Total= $9.60
3. $11.79
4. $195.83
5.

Selling Price	Tax Rate	Quantity of Tax	Total Cost
$400.00	12%(GST+PST)	$48.00	$448.00
$550.00	GST(5%)	$27.50	$577.50
$800.00	PST(7%)	$56	$856
$590.00	15%	$88.50	$678.50
$58.82	17%	$10.00	$68.82
$208.93	12%	$25.07	$234.00
$200	15%	$30.00	$230

Level B

1. $ 2889.60
2. $ 6718.88
3. $ 111.99
4. $ 809.99
5. $ 33.59
6.

Regular Price	Discount Rate	Amount of Discount	Tax Rate	Amount of Tax	Sale Price
$ 259.99	10%	$ 26.00	GST+PST(12%)	$ 28.08	$ 262.07
$ 99.99	15%	$ 15.00	10%	$ 8.50	$ 93.49
$ 300.00	0	0	12%	$ 36.00	$ 336.00
$ 599.99	20%	$ 120.00	3.25%	$ 15.60	$ 495.59
$ 199.99	12%	$ 24.00	10%	$ 17.60	$ 193.59
$ 1350.00	0	0	12%	$ 162.00	$ 1512

7. Original Price= $ 1449.78; tax= $ 139.18
8. 5%
9. Original Price= $ 510; tax= $ 45.90

Level C

1. x: price of a chair

$x+10$: price of a table

$\frac{x}{x+10}=60\%$

$x=60\%(x+10)=0.6(x+10)=0.6x+6$

$x-0.6x=6$

$0.4x=6$

$x=\frac{6}{0.4}=15$

chair: $ 15, table: $ 25

2. $\frac{350-280}{700}=\frac{70}{700}=10\%$ (increase)

3. 5% of 80 g salt water has salt:

$5\%\times 80=4$ g

8% of 20 g salt water has salt:

$8\%\times 20=1.6$ g

So the mixed solution has 5.6 g salt.

The total solution is $80+20=100$ g.

10 g out of 100 g solution is poured away.

That is $\frac{10}{100}=10\%$, then the salt left is

$5.6\times(1-10\%)=5.6\times 0.9=5.04$ g.

The new concentration is $\frac{5.04}{100-10+10}=$

5.04%

4. x: the wholesale price of the set of teacups

$x(1+20\%)(1+7\%+5\%)(1-15\%)=71.4$

$x(1.2\times 1.12\times 0.85)=71.4$

$x=62.5$

2.4 Answer Key

Level A

1.

Total Sale	Rate of Commission	Amount of Commission
$ 265 000	2%	$ 5300
$ 6500	3%	$ 195
$ 1 659 800	2%	$ 33 196
$ 239 850	0.85%	$ 2038.73
$ 565 800	3%	$ 16 974
$ 7375	4%	$ 295
$ 294 800	2.5%	$ 7370
$ 39 810	1.85%	$ 736.49

2.

Principal	Interest Rate Per Year	Years	Simple Interest	Total Amount
$ 5000.00	5%	2	$ 500.00	$ 5500.00
$ 45 500.00	3%	3	$ 4095.00	$ 49 595.00
$ 2650.00	4%	4	$ 424.00	$ 3074.00
$ 9800.00	15%	2	$ 2940.00	$ 12 740.00
$ 236 500.00	8%	5	$ 94 600.00	$ 331 100.00
$ 39 800.00	5%	3	$ 5970.00	$ 45 770.00
$ 1225.81	12%	2	$ 294.19	$ 1520.00
$ 34 500.00	6%	5	$ 10 350.00	$ 44 850.00

Level B

1. B: $ 2550.00; $ 5350.00

C: $ 315.00; $ 3200.00

D: $ 5870.00; $ 117.40

E: $2280.00; $4780.00

F: $375.90; $2880.00

G: $65 200.00; $1304.00

2. Mr. Li

3. If Bill sells less than $16666.67, 2% commission is better.

If Bill sells more than $16666.67, 5% commission is better.

4.

Original Number	New Number After Changing	Amount of Change	Percent of Change
$5000	$5500	+$500	+10%
$4800	$5400	+$600	+12.5%
$5900	$5100	−$800	−13.56%
$300	$350	+$50	+16.67%
$2800	$2660	−$140	−5%
$6200	$4600	−$1600	−25.81%
$600	$640	+$40	+6.67%

5.

Principal	Interest Rate Per Year	Years	Compound Interest	Total Amount
$5000	5%	1	$250	$5250
$4500	3%	2	$274.05	$4774.05
$8500	4%	1	$340	$8840
$1000	15%	2	$322.5	$1322.50
$2500	8%	2	$416	$2916
$3900	5%	2	$399.75	$4299.75
$4800	12%	2	$1221.12	$6021.12

6.

Year	$ at start of year	Interest	$ at end of year
1	25 000	2000	27 000
2	27 000	2160	29 160
3	20 160	2332.80	31 492.80
4	31 492.80	2519.42	34 012.22
5	34 012.22	2720.98	36 733.20

Level C

1. x: number of boys

$x(1+12\%)$: number of girls

$x+x(1+12\%)=318$

$x(1+1+12\%)=318$

$x=\frac{318}{2.12}=150$

female students: $150\times 1.12=168$

2. The number of apples in Box A: x

The number of apples in Box B: $\frac{1}{2}x$

$\frac{\frac{1}{2}x+25}{x-25}=80\%$

$0.5x+25=0.8(x-25)$

$=0.8x-20$

$25+20=0.3x$

$x=\frac{45}{0.3}=150$

$x-25=125$

Box A has 125 apples now.

3. The number of milk candies: x

The number of total candies: $\frac{x}{45\%}$

$\frac{x}{\frac{x}{45\%}+16}=25\%$

$x=25\%\left(\frac{x}{45\%}+16\right)$

$=0.25\left(\frac{x}{0.45}+16\right)$

$=\frac{0.25}{0.45}x+4$

$=\frac{5}{9}x+4$

$\left(1-\frac{5}{9}\right)x=4$

$\frac{4}{9}x=4$

$x=9$

9 milk candies.

4. Chris' interest income in 4 years:

$250\ 000\times(1+1.15\%)^4-250\ 000$

$=11\ 699.90$

Chris' salary income in 4 years:

$2000\times 12\times 4=96\ 000$

Chris' commission income in 4 years:

$7\times 14\ 000\times 3.5\%\times 12\times 4=164\ 640$

His total income:

$11\ 699.90+96\ 000+164\ 640=\$272\ 339.90$

2.5 Answer Key

Level A

a) 11 000 ppm; 11 000 000 ppb

b) 5000 ppm; 5 000 000 ppb

c) 1 000 000 ppm; 1 000 000 000 ppb

d) 199 000 ppm; 199 000 000 ppb

e) 70 000 ppm; 70 000 000 ppb

f) 35 000 ppm; 35 000 000 ppb

g) 300 ppm; 300 000 ppb

h) 20 000 ppm; 20 000 000 ppb

i) 10 300 ppm; 10 300 000 ppb

j) 8810 ppm; 8 810 000 ppb

Level B

a) 0.0025%; 0.025‰

b) 0.0002%; 0.002‰

c) 0.00005%; 0.0005‰

d) 0.075%; 0.75‰

e) 1.5%; 15‰

f) 0.000058%; 0.00058‰

g) 0.0000029%; 0.000029‰

h) 0.0000003%; 0.000003‰

i) 0.00000004%; 0.0000004‰

j) 0.0006%; 0.006‰

Level C

$$\frac{0.01\ \text{mg/m}^3}{8.65\ \text{mg/cm}^3}=\frac{\frac{0.01\ \text{mg}}{\text{m}^3}\cdot\left(\frac{1\ \text{m}}{100\ \text{cm}}\right)^3}{8.65\ \text{mg/cm}^3}$$

$$=\frac{\frac{0.01}{100^3}\ \text{mg/cm}^3}{8.65\ \text{mg/cm}^3}=\frac{10^{-8}}{8.65}=1.2\times10^{-9}$$

$$=1.2\ \text{ppb}$$

Chapter 2 TEST

1. a) 42.86% b) 33.33%
   c) 18.75% d) 62.5%
   e) 46.875% f) 225%
   g) 23.33% h) 25%

2. a) $2\frac{659}{1000}$ b) $\frac{3}{8}$
   c) $15\frac{1}{5}$ d) $20\frac{1}{20}$
   e) $1\frac{266}{625}$ f) $\frac{27}{125}$
   g) $2\frac{1677}{10\ 000}$ h) $6\frac{5923}{10\ 000}$

3. 83.5%

4. 12.5%

5. a) 12.8% b) 43.94%
   c) 1440 d) 0.018
   e) 6580 f) 539.5

6. $355\frac{5}{9}$

7. 45.88

8. a) 40 b) 5.53
   c) 1.12 d) 4.65

9.

Regular Price	Discount Rate	Amount of Discount	Tax Rate	Amount of Tax	Sale Price
\$49.99	10%	\$5.00	GST+ PST (12%)	\$5.4	\$50.39
\$199.99	15%	\$30	7%	\$11.9	\$181.89
\$200.00	0%	0	7%	\$14.00	\$214
\$499.99	40%	\$199.99	8.3%	\$25.00	\$325

10. Original Price: \$799.99; Tax: \$72.00

11. Science: \$113.33; Math: \$133.33

12.

Principal	Interest Rate Per Year	Years	Simple Interest	Total Amount
\$45 500.00	3%	3	\$4095.00	\$49 595.00
\$2650.00	4%	4	\$424.00	\$3074.00
\$236 500.00	8%	5	\$94 600.00	\$331 100.00
\$34 500.00	6%	5	\$10 350.00	\$44 850.00

13. A: \$1125.00
    B: \$2550.00; \$5350.00
    C: \$2280.00; \$4780.00
    D: \$65 200.00; \$1304.00

14.

Principal	Interest Aate Per Year	Years	Compound Interest	Total Amount
\$2500	5%	1	\$125	\$2625
\$8200	4.5%	2	\$754.61	\$8954.61
\$3200	8%	1	\$256	\$3456
\$4800	12%	2	\$1221.12	\$6021.12

ANSWERS CHAPTER 3

3.1 Answer Key

Level A

1. Answers vary, test with calculator

2. a) 24 b) 45
   c) 128.3 d) 114
   e) 40 f) 106.7
   g) 112 h) 40.3
   i) 17.2 j) 133.6
   k) 27.6 l) 141.7

Level B

1. Answers vary, test with calculator

2. a) $x=2, y=10$ b) $x=3.4, y=9.4$
   c) $x=4.1, y=275$ d) $x=8.3, y=4.3$
   e) $x=75, y=18$ f) $x=70, y=44.5$
   g) $x=120, y=6.3$ h) $x=125, y=160$
   i) $x=191.7, y=703.0$

3. a) $x=24, y=21$ b) $x=22.3, y=21$
   c) $x=30.7, y=26.0$ d) $x=2.7, y=24$
   e) $x=69, y=168$ f) $x=54, y=15$
   g) $x=1.7, y=45$

4. a) 17 : 35 b) 18 : 35 c) 18 : 17

5. 350 km

6. 91 m^3

7. 16 000 bricks

8. 48 minutes

9. a) 15 : 4 b) 15 : 14

10. a) 63 b) 50

Level C

1. Assume each container can carry L.

   Salt in the first container is $\frac{2}{2+3}L=\frac{2}{5}L$.

   Salt in the second container is $\frac{3}{3+4}L=\frac{3}{7}L$.

   The ratio of salt to water in the newly mixed salt water is:

   $\frac{2}{5}L+\frac{3}{7}L : \frac{3}{5}L+\frac{4}{7}L$

   $=\frac{14+15}{35}L : \frac{21+20}{35}L=29 : 41$

2. Bag A's weight: $4x$

   Bag B's weight: x

   $4x-10 : x+10=7 : 5$

   $7(x+10)=5(4x-10)$

   $7x+70=20x-50$

   $120=13x$

   $x=\frac{120}{13}$

   combined weight $=4x+x=5x=5\times\frac{120}{13}=$

   $46\frac{2}{13}$ g

3.

	Daily working speed	Weekly working speed
Larry	$\frac{1}{40}$	$\frac{1}{40}\times5=\frac{1}{8}$
Moe	$\frac{1}{35}$	$\frac{1}{35}\times4=\frac{4}{35}$
Harry	$\frac{1}{50}$	$\frac{1}{50}\times3=\frac{3}{50}$

   From the table above, their combined weekly working speed is:

$$\frac{1}{8}+\frac{4}{35}+\frac{3}{50}$$

   time:

$$\frac{1}{\frac{1}{8}+\frac{4}{35}+\frac{3}{50}}=4$$

4. a) x: the amount of butter

   y: the amount of sugar

   z: the amount of milk

   $x : 3 : y : z=2 : 4 : 1\frac{1}{2} : \frac{3}{4}$

   $x : 3=2 : 4, x=\frac{2\times3}{4}=\frac{3}{2}=1.5$

   $3 : y=4 : 1\frac{1}{2}, y=\frac{3\times1\frac{1}{2}}{4}=\frac{9}{8}$

$$3 : z = 4 : \frac{3}{4}, z = \frac{3\times\frac{3}{4}}{4} = \frac{9}{16}$$

b) $\frac{9}{8}\times\frac{200}{1}\times\frac{1}{454}\times\frac{1}{5}\times\frac{\$2.82}{1}$

$$= \frac{9\times200\times2.82}{8\times454\times5} = 28$$

3.2 Answer Key

Level A

1. a) 600 pages/box b) 7.1 calls/day
   c) 60 km/hr d) 8.3 cal/min
   e) 60 words/min f) $2600/month
   g) 4 slices/person h) 16.7 clients/worker
   i) $850/suit j) 4 acres/tank
2. a) 24 kg b) $175
   c) 966.67 km d) 10 980 m
   e) 22 500 words f) 6.90%

Level B

1. a) 0.6 m/hr b) 30 hrs
   c) 40 hrs d) 5 days
2.

Name	Mary	Tony	Arleen	Martin	Wilson
Quantity of typed words	1000	1000	1000	1000	1000
Time taken in minutes	25	25	30	25	23.9
Quantity of wrong words	25	250	100	0	210
Quantity of correct words	975	750	900	1000	790
Speed (words/min)	39	30	30	40	33
Order in the contest	2	4	4	1	3

3. 27, 21
4. 1 USD to 1.0102 CAD
5. a) 4088 m b) 6.85 s
6. 1 USD to 1.0102 CAD
   1 CAD to 0.9899 USD
7. B; B; A; A; A
8. Cement: 19.2 tons; Sand: 28.8 tons; Stones: 48 tons
9. 21 cm; 28 cm; 35 cm
10. 44, 77, 143
11. a) 12 g b) 450 g c) 6448 g
12. 13.89 km/hr
13. 21.93 mins

Level C

1. x: the number of people in the singing club
   y: the number of people in the dancing club

$$\begin{cases} x : y = 3 : 2 \\ x-10 : y+10 = 7 : 8 \end{cases}$$

$$\rightarrow \begin{cases} 2x = 3y \\ 8(x-10) = 7(y+10) \end{cases}$$

$$\rightarrow \begin{cases} x = \frac{3y}{2} & ① \\ 8x-80 = 7y+70 & ② \end{cases}$$

Substitute ① into ②.

$$8\left(\frac{3y}{2}\right)-80 = 7y+70$$

$$12y-80 = 7y+70$$

$$5y = 150$$

$$y = 30$$

$$x = 45$$

So, there are 45 people in the singing club.

2. x: merchandise stored in Warehouse A
   y: merchandise stored in Warehouse B

$$\begin{cases} x : y = 4 : 3 \\ x-8 : y+8 = 4 : 5 \end{cases} \rightarrow \begin{cases} 3x = 4y \\ 5(x-8) = 4(y+8) \end{cases}$$

$$\rightarrow \begin{cases} \frac{3x}{4} = y & ① \\ 5X-40 = 4y+32 & ② \end{cases}$$

Substitute ① into ②.

$$5x-40 = 4\times\frac{3x}{4}+32$$

$5x-40=3x+32$

$2x=72$

$x=36, y=27$

3. Assume Kevin's walking speed is 7 km/h, Robin's walking speed is 5 km/h.

Then the distance between C and D (d) is:

$d=7\times0.5+5\times0.5=6$ km

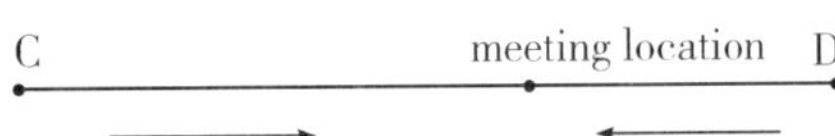

In the second case, assume it takes Kevin t hours to catch up with Robin.

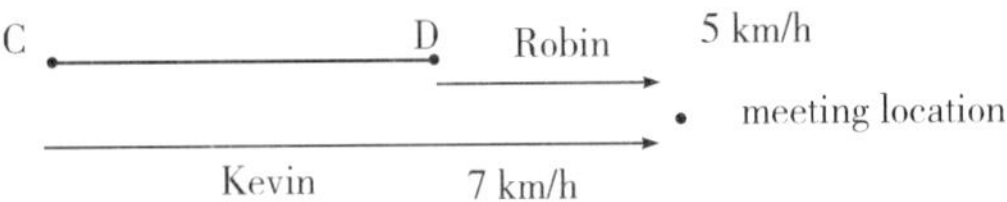

Then the equation is

$7t=6+5t$

$2t=6$

$t=3$ hours

4. Daniel eats $\frac{5}{3}$ cookies/day.

Dennis eats $\frac{10}{7}$ cookies/day.

Donna eats $\frac{1}{2}$ cookies/day.

So together they can eat $(\frac{5}{3}+\frac{10}{7}+\frac{1}{2})$ cookies/day.

$$\frac{50\times2}{\left(\frac{5}{3}+\frac{10}{7}+\frac{1}{2}\right)}=27.8=28 \text{ days}$$

3.3 Answer Key

Level A

1. 368.42 km
2. 2.73 cm
3. 75
4. 27 000
5. 85.32 km/hr

Level B

1. 65
2. 18 500 calories
3. 25.47 km
4. 355.2 miles
5. 153 pages

Level C

1. $\frac{14\ 000}{2800\times30\%}=16.7\approx17$ months
2. $\frac{149\ 597\ 871}{300\ 000}=498.66$ s

Chapter 3 TEST

1. a) 14 b) 44
   c) 78 d) 57
   e) 82.8 f) 281.3
2. a) $x=4, y=2.7$ b) $x=2, y=8$
   c) $x=12.9, y=126$ d) $x=5.3, y=6$
   e) $x=69, y=138$ f) $x=45, y=30$
   g) $x=5, y=30$
3. a) 5 : 9 b) 4 : 9 c) 4 : 5
4. 49 km
5. a) 70 words/min b) 5 slices/person
   c) \$ 2100/month d) 20 clients/worker
   e) \$ 890/suit f) 4 acres/tank
6. a) 336 kg of sand b) \$ 221
   c) 592 km d) 11 250 m
7. a) $\frac{8}{11}$ m/hr b) $19\frac{1}{4}$ hr
   c) 27 hrs 30 mins d) 3 days
8.

Name	Quantity of Typed Words	Time Taken in Minutes	Quantity of Wrong Words	Quantity of Correct Words	Speed (words/min)	Order in the Contest
Jerry	1500	20	20	1480	74	2
Cindy	1500	20	125	1375	68.75	3
Linda	1500	45.83	125	1375	30	5
Alyssa	1500	20	0	1500	75	1
Kevin	1500	40	180	1320	33	4

9. 38 34
10. 1 USD to 1.0515 CAD
    1 CAD to 0.9510 USD
11. 1 004 310
12. a) 6.5 hr b) 52.3 L

ANSWERS CHAPTER 4

4.1 Answer Key

Level A

1. a) $\frac{1}{77}$ b) $\frac{1}{9}$ c) $\frac{1}{21}$

   d) $\frac{3}{4}$ e) $\frac{2}{21}$ f) $\frac{7}{88}$

   g) $\frac{1}{14}$ h) $\frac{1}{7}$ i) $\frac{1}{30}$

2. a) $16\frac{1}{10}$ b) $6\frac{3}{4}$ c) $11\frac{9}{10}$

   d) $14\frac{5}{8}$ e) 16 f) $16\frac{13}{15}$

   g) $24\frac{4}{9}$ h) $19\frac{1}{2}$ i) 15

Level B

1. a) $\frac{9}{13}$ b) $\frac{1}{42}$ c) $\frac{2}{9}$

   d) $\frac{1}{42}$ e) $\frac{1}{6}$ f) $\frac{1}{9}$

   g) $\frac{1}{8}$ h) $\frac{11}{13}$ i) $\frac{11}{26}$

2. a) $\frac{5}{8}$ b) 118.75 mL

3. 35 students/bus

4. a) $\frac{2}{3}$ b) $\frac{1}{8}$ c) $\frac{3}{8}$

   d) $\frac{1}{36}$ e) $\frac{48}{49}$ f) $\frac{3}{22}$

   g) $\frac{1}{21}$ h) $2\frac{1}{3}$ i) $\frac{2}{7}$

5. a) $\frac{14}{19}$ b) $1\frac{5}{51}$ c) $\frac{31}{42}$

   d) $\frac{26}{27}$ e) $1\frac{5}{6}$ f) $1\frac{3}{11}$

   g) $\frac{24}{35}$ h) $\frac{14}{27}$ i) $\frac{69}{91}$

6. 40, 30

7. 88

8. 62 km/h

Level C

1. First day: $\frac{1}{3}$

   second day: $\frac{1}{3}$

   Third day: $\frac{1}{4}\times(1-\frac{1}{3}-\frac{1}{3})=\frac{1}{12}$

   Fourth day: $\frac{1}{2}\times(1-\frac{1}{3}-\frac{1}{3}-\frac{1}{12})=\frac{1}{8}$

   $\frac{1}{8}\times 70\ 000=\$8750$

2. a) The number of total possible tickets is $_{38}C_5=501\ 942$

   The number of total tickets they bought:

   $150+150+\frac{9}{2}\times(150+150)=1650$

   The probability is: $\frac{1650}{501\ 942}=\frac{825}{250\ 971}$

   b) Tony's share of the jackpot:

   $50\ 000\times\frac{\frac{9}{2}(150+150)}{1650}=40\ 909.09$

   His profit is:

   $40\ 909.09-\frac{9}{2}(150+150)\times 2.5$

   $=\$37\ 534.09=\$37\ 534$

4.2 Answer Key

Level A

1. a) +12 b) +18 c) +21

   d) +12 e) −15 f) −24

   g) −28 h) −16 i) −24

   j) −24 k) −24 l) −20

   m) −16 n) −30 o) −40

   p) −27 q) +20 r) +54

   s) −42 t) −32 u) +20

2. a) +2 b) +3 c) +5

   d) +2 e) −2 f) −4

   g) −3 h) −4 i) −2

   j) −2 k) −4 l) −5

   m) −4 n) −3 o) +2

   p) −3 q) −4 r) +2

   s) +2 t) −6 u) +8

Level B

1. a) +546 b) +575 c) +714
   d) +495 e) −1173 f) −806
   g) −882 h) −1281 i) −868
   j) −1932 k) −5712 l) +5928
   m) −22176 n) +215 o) −129
   p) −116 q) +76
2. a) +21 b) +16 c) +32
   d) +16 e) −21 f) −15
   g) −32 h) −15 i) −3
   j) +3 k) +1 l) +2
   m) −3 n) +96 o) −28
   p) −187 q) +126
3. 143.5 cm
4. 27.8℃
5. 12 pupils
6. $20

Level C

1. Johnny's mark: $65\times(1+5\%)=68.25$
   Chris' mark: $68.25\times(1+15\%)=78.4875$
   Eric's mark: $\frac{68.25+78.4875}{2}\times(1-10\%)$
   $=66.03$
   Eric passed the course.
2. The snail climbs up totally 2 m for every three days.
   It takes the snail $\frac{12}{2}\times3=18$ day to reach 12 m high. Then it takes one more day to go another 3 m to reach the top.
   So totally it takes 19 days.

4.3 Answer Key

Level A

a) $3\frac{1}{3}$ b) $5\frac{1}{4}$ c) $4\frac{2}{3}$

d) 8 e) $2\frac{21}{32}$ f) $2\frac{2}{3}$

g) 11 h) $4\frac{4}{9}$ i) 3

j) −6 k) −3 l) $-3\frac{19}{24}$

m) −10 n) 14 o) $3\frac{1}{3}$

Level B

1. a) 4 b) $\frac{7}{12}$ c) $\frac{1}{3}$
   d) $1\frac{1}{2}$ e) $5\frac{2}{5}$ f) $\frac{3}{5}$
   g) $\frac{4}{5}$ h) $\frac{1}{2}$ i) $-\frac{1}{2}$
   j) $-1\frac{1}{2}$ k) $-\frac{9}{34}$ l) $-\frac{5}{9}$
   m) $-1\frac{7}{9}$ n) $-\frac{8}{21}$ o) $-\frac{4}{5}$
2. a) $1\frac{1}{2}$ b) 3 c) $1\frac{143}{432}$
   d) $3\frac{1}{5}$ e) $-1\frac{27}{64}$ f) $-9\frac{9}{10}$
   g) $-\frac{5}{14}$ h) $12\frac{51}{112}$ i) $-1\frac{27}{28}$
   j) $10\frac{85}{224}$
3. a) $51\frac{61}{324}$ b) $-7\frac{1}{6}$ c) $-1\frac{11}{21}$

Level C

1. A: Abraham's age
   B: Bart's age (15)
   H: Homer's age

$$\begin{cases}(B-5)=\frac{1}{4}(A-5)\\(B+15)\times1.5=H+15\end{cases}$$

$$\begin{cases}15-5=\frac{1}{4}(A-5)\\(15+15)\times1.5=H+15\end{cases}$$

$$\begin{cases}10=\frac{1}{4}(A-5)\\45=H+15\end{cases}\longrightarrow A=45, B=30$$

The age difference is $A-B=15$.

2.

	Hawaiian pizza	Greek Pizza
end of day 1	$1-\frac{3}{8}=\frac{5}{8}$	1
end of day 2	$\frac{5}{8}-\frac{5}{8}\times\frac{1}{2}=\frac{5}{16}$	$1-\frac{2}{3}=\frac{1}{3}$
end of day 3	$\frac{5}{16}$	$\frac{1}{3}-\frac{1}{3}\times\frac{1}{2}=\frac{1}{6}$

Chapter 4 TEST

1. a) $\frac{3}{4}$ b) $\frac{2}{21}$ c) $\frac{7}{88}$
   d) $\frac{3}{7}$ e) $\frac{1}{7}$ f) $\frac{1}{36}$
2. a) $11\frac{2}{3}$ b) 9 c) 15
   d) $19\frac{1}{4}$ e) 21 f) 18
3. a) $\frac{1}{45}$ b) $\frac{10}{21}$ c) $\frac{1}{24}$
   d) $\frac{1}{6}$ e) $\frac{11}{13}$ f) $\frac{8}{15}$
4. a) $\frac{13}{588}$ b) $\frac{48}{49}$ c) $\frac{32}{99}$
   d) $\frac{1}{19}$ e) $4\frac{2}{3}$ f) $\frac{11}{35}$
5. a) -805 b) 1426 c) 448
   d) -646 e) -783 f) 1554
6. a) -36 b) -36 c) -27
   d) -35 e) -19 f) -27
7. a) $-7\frac{1}{5}$ b) $-3\frac{9}{11}$ c) -3
   d) $-4\frac{4}{9}$ e) 7 f) $20\frac{2}{3}$
8. a) $-\frac{7}{9}$ b) $2\frac{3}{5}$ c) $-1\frac{1}{3}$
   d) $-1\frac{1}{2}$ e) $\frac{64}{105}$ f) $-\frac{7}{8}$
9. a) $2\frac{13}{16}$ b) $2\frac{2}{3}$ c) $-\frac{8}{63}$ d) 18
10. a) 131 b) $-13\frac{3}{10}$ c) $-7\frac{7}{15}$
11. 24°
12. \$40
13. a) $\frac{1}{3}$ b) 300 mL
14. $87\frac{1}{2}$ $7\frac{1}{2}$
15. 68 km/h
16. $\frac{2009}{360}$ km

ANSWERS CHAPTER 5

5.1 Answer Key

Level A

1. a)

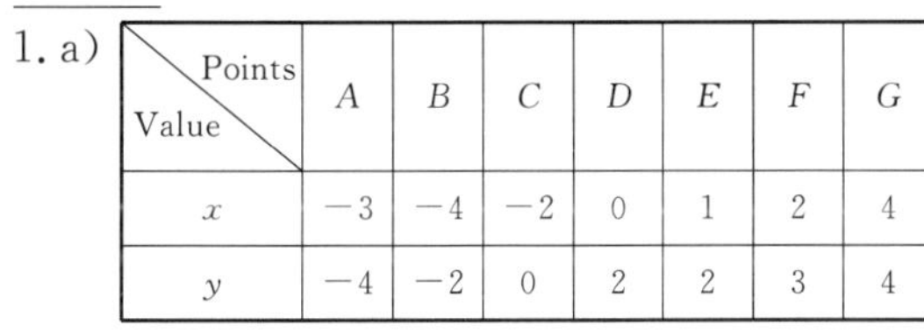

Value \ Points	A	B	C	D	E	F	G
x	-3	-4	-2	0	1	2	4
y	-4	-2	0	2	2	3	4

b)

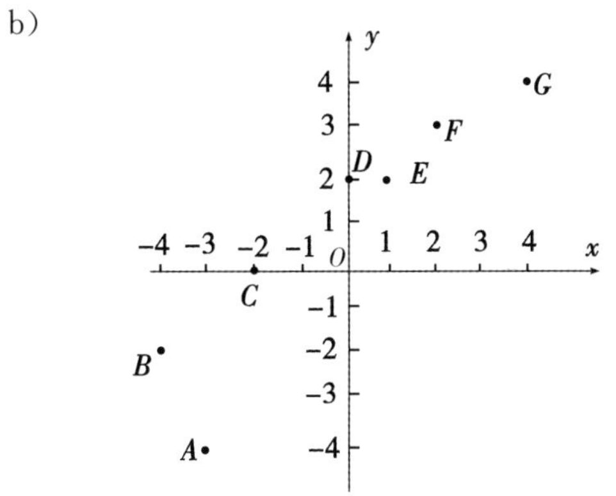

2. a) (3, 2) b) (2, 4) c) (3, 4)
   d) (4, 2) e) (4, 5) f) (3, 1)
   g) (−3, 2) h) (2, −2) i) (−2, −5)
   j) (−4, 2) k) (−3, −2) l) (3, −2)

Value \ Points	A	B	C	D	E	F	G	H	I	J	K	L
x	3	2	3	4	4	3	-3	2	-2	-4	-3	3
y	2	4	4	2	5	1	2	-2	-5	2	-2	-2

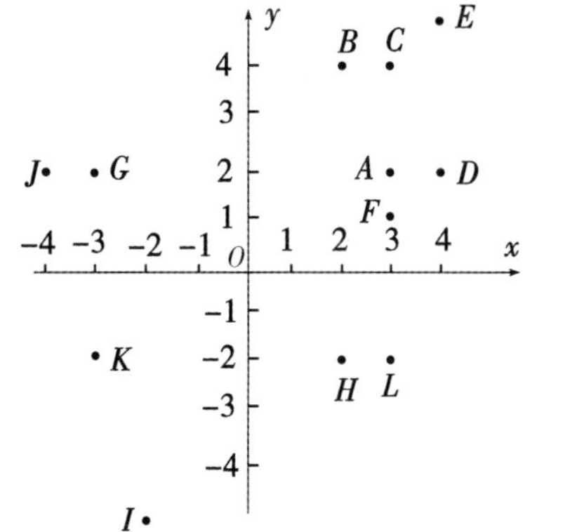

Level B

1. a) 21 b) 27 c) -12

d) $\frac{1}{8}$ e) $-\frac{4}{5}$ f) $-\frac{1}{40}$

g) $\frac{1}{24}$ h) $-\frac{1}{8}$ i) 19.5

j) $\frac{7}{8}$ k) $-\frac{5}{48}$ l) $-\frac{18}{17}$

2. a) 9, 10, 11 (equation is $y=x+3$)

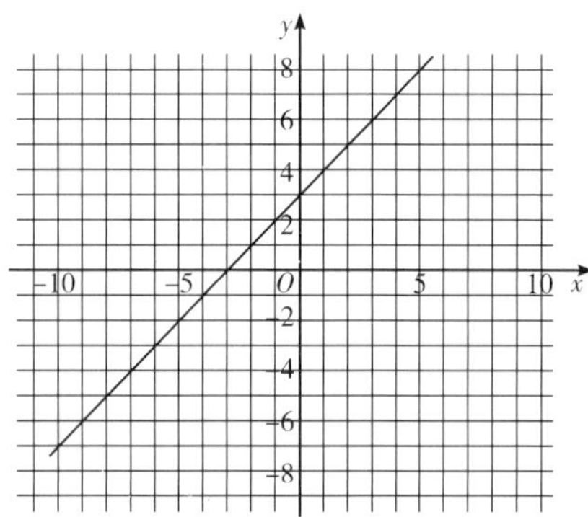

b) 4, 3, 2 (equation is $y=-x+10$)

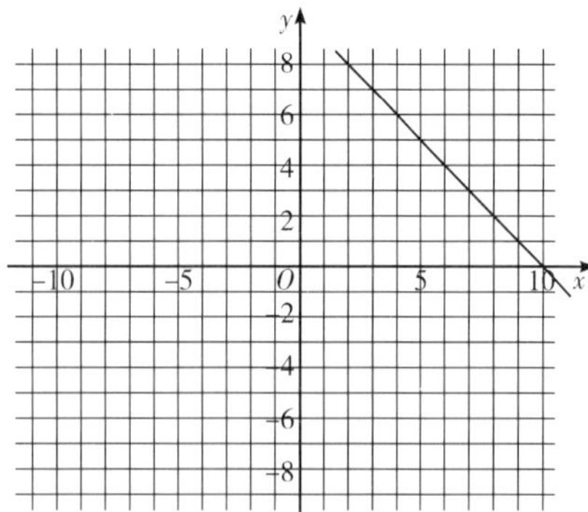

c) 6, 6, 6 (equation is $y=6$)

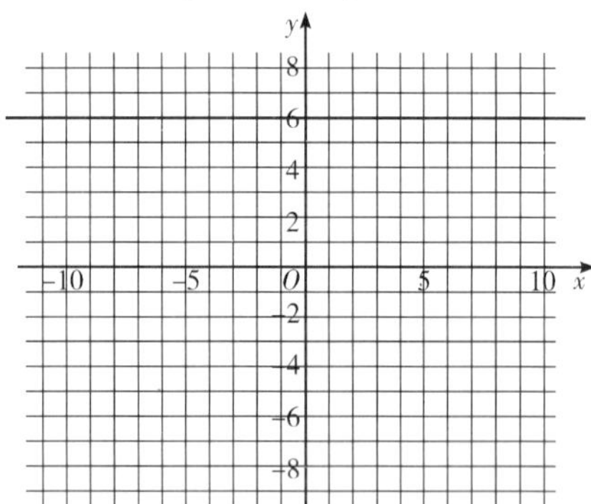

Level C

1. B: base pay

D: sold price of one doughnut

Jim: $135=B+400D$ .... ①

Harry: $105=B+300D$ .... ②

Michael: $225=B+700D$ .... ③

① − ②:

$30=100D$, $D=0.3$, $B=15$

a) each doughnut sold: \$0.3

b) base pay: \$15

c) $B+550D$

$=15+550\times 0.3$

$=\$180$

2. a) From 9 am to noon, $250-175=75$ were sold. $\frac{75}{3\text{ h}}=25/\text{h}$

b) At 8 am, there are $250+25=275$ cupcakes.

c) At 3 pm there are 100 left.

$\frac{100}{25}=4$ h

It takes 4 more hours, so at 7 pm the cupcakes will be sold out.

5.2 Answer Key

Level A

1. a)

x	3	4	5	6
y	4	5	6	7

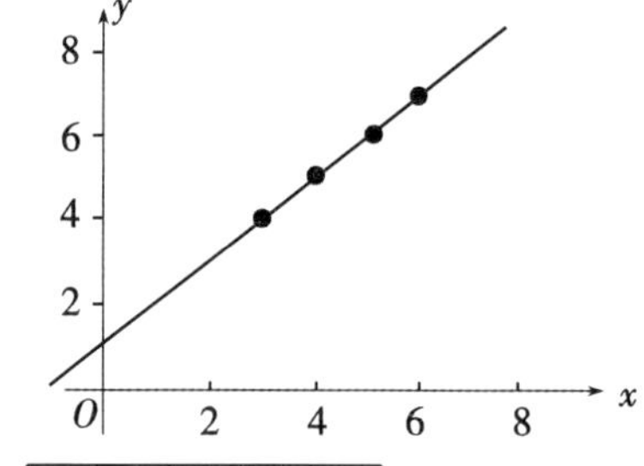

b)

x	−4	−3	−2	−1
y	−6	−4	−2	0

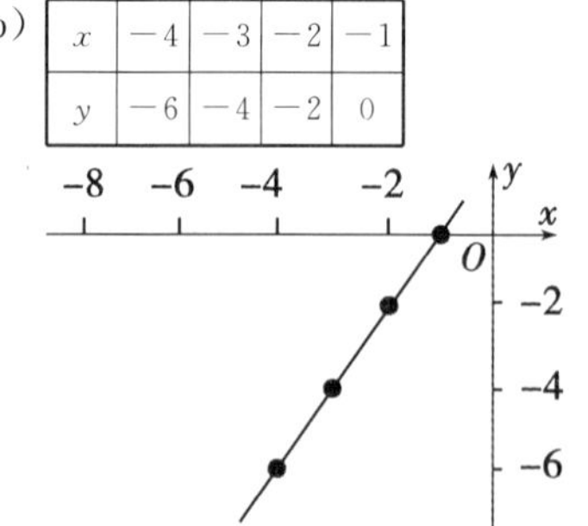

c)

x	1	2	3	4
y	7	5	3	1

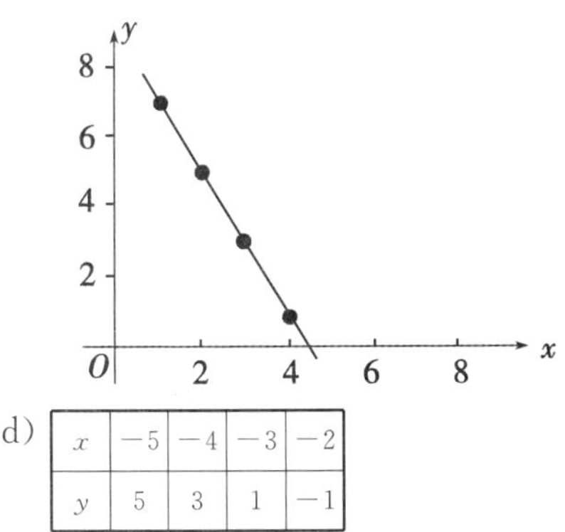

d)

x	-5	-4	-3	-2
y	5	3	1	-1

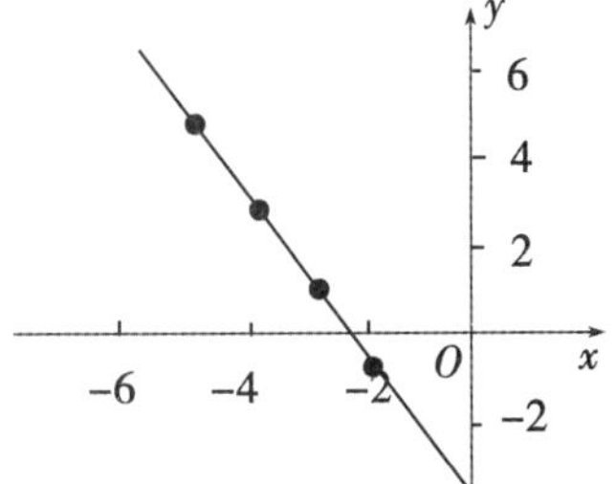

e)

x	-2	-1	0	1
y	3	4	5	6

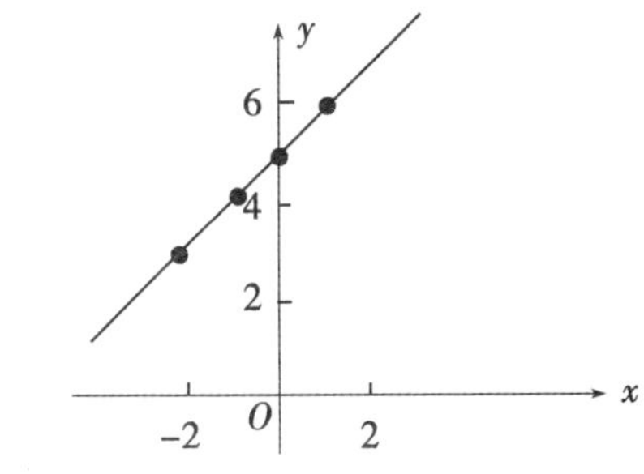

f)

x	0	1	2	3
y	0	1	4	9

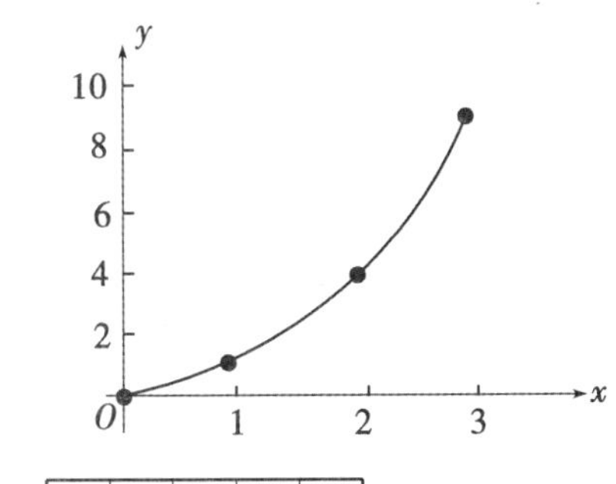

g)

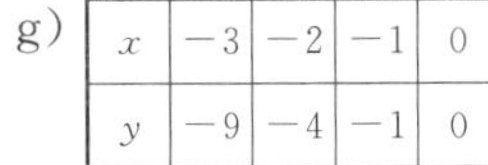

x	-3	-2	-1	0
y	-9	-4	-1	0

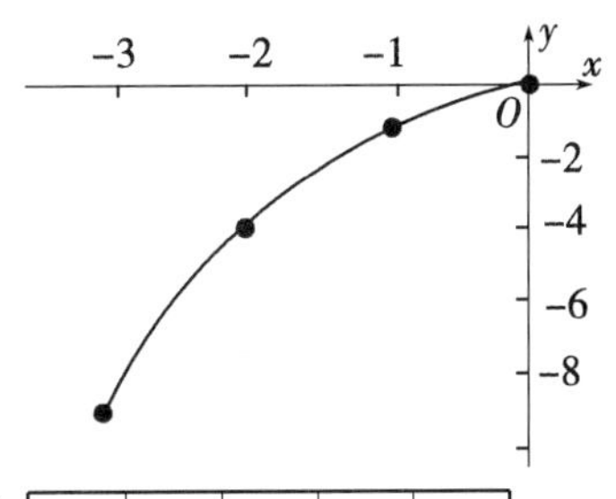

h)

x	-3	-2	-1	1
y	-6	-4	-2	2

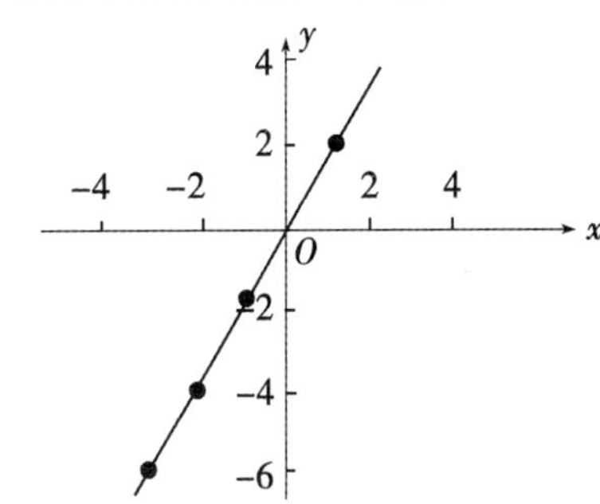

2. a) 6, 8 (equation is $y=2x+4$)

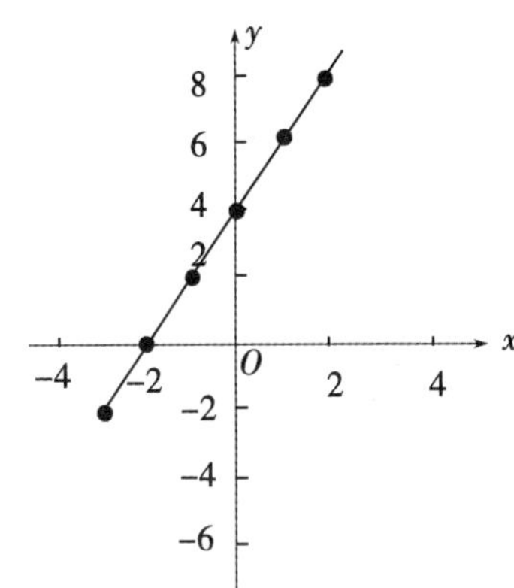

b) -3, -5, -7 (equation is $y=-2x-1$)

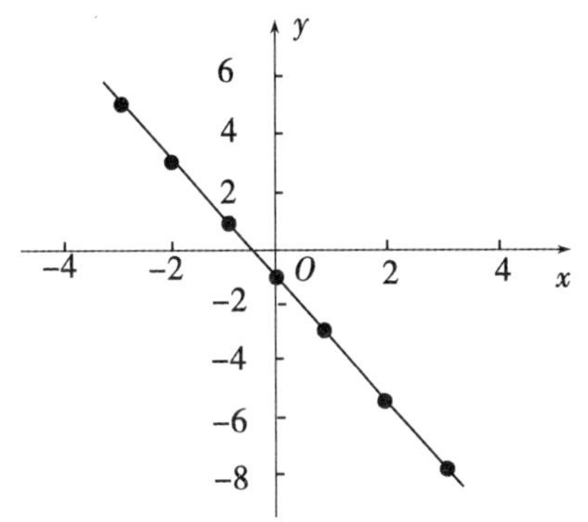

Level B

1. a)

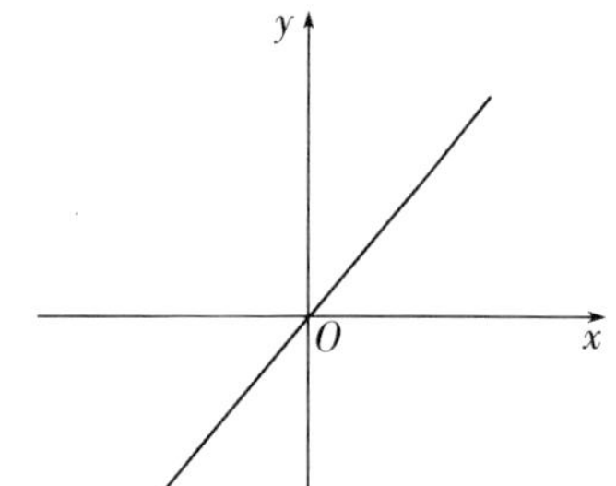

b)

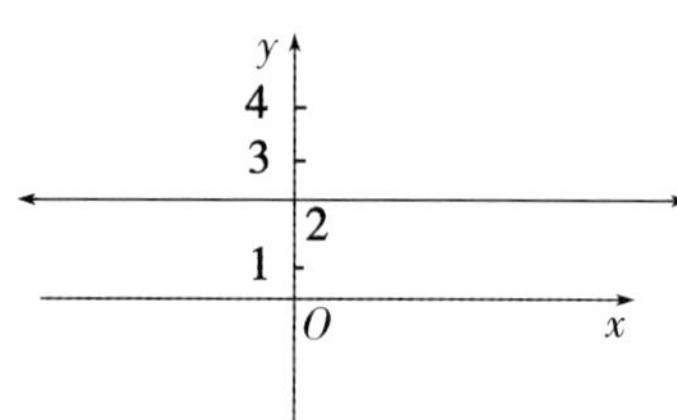

c)

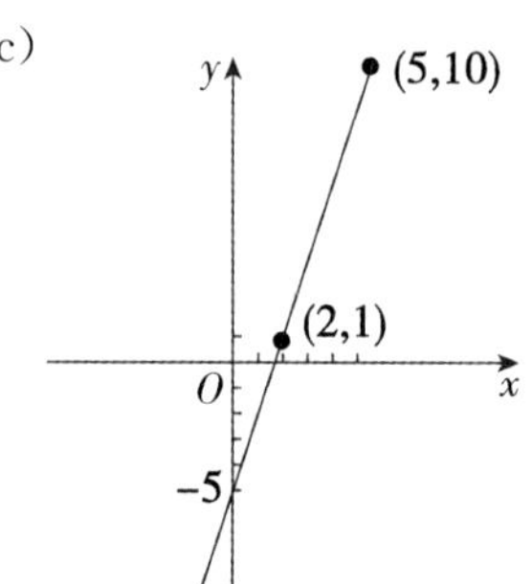

d)

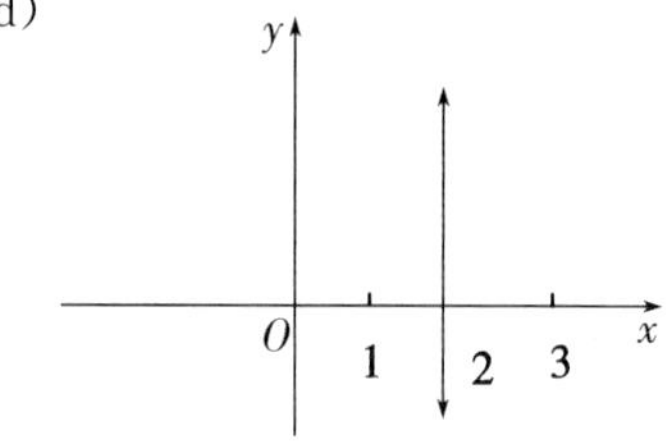

e)

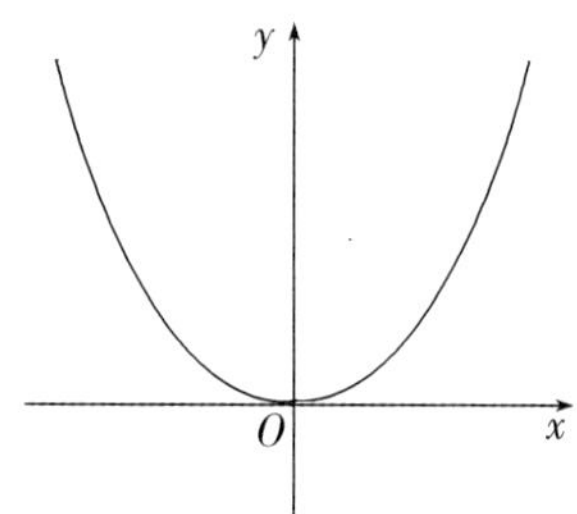

f)

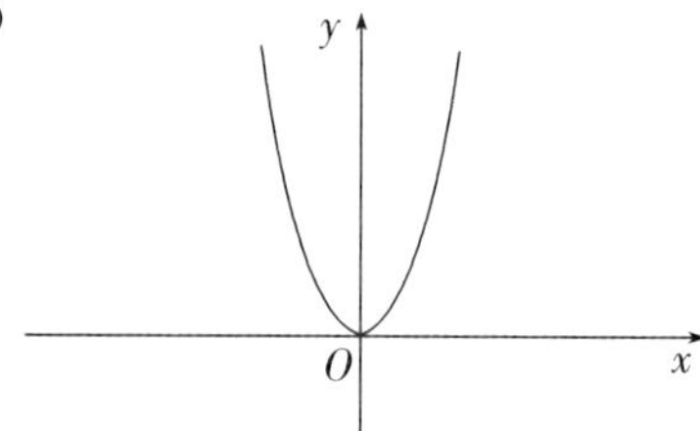

g)

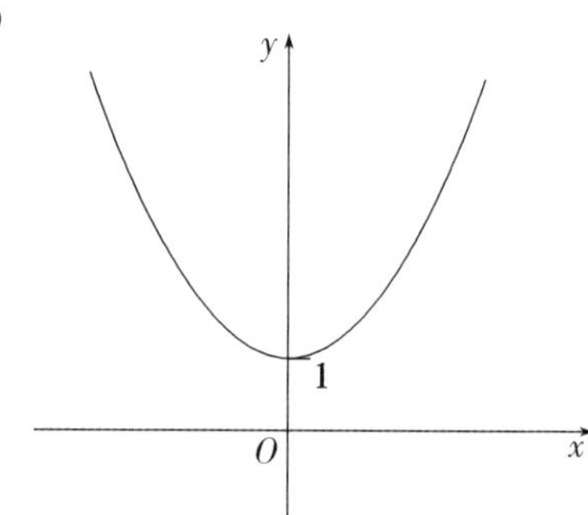

h)

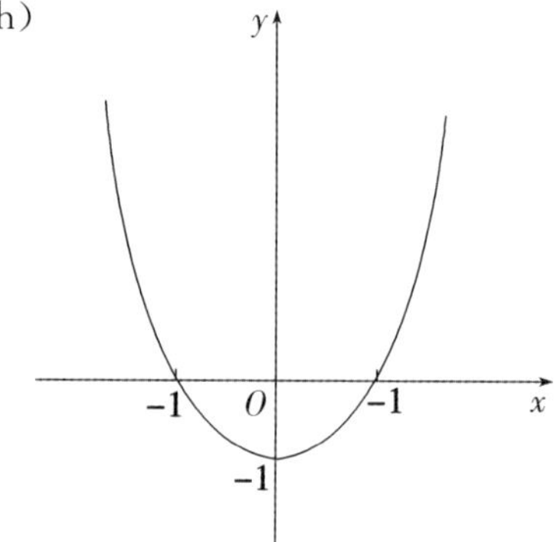

2. a) 36, 42 (equation is $y=6x$)

   b) -11, -14 (equation is $y=-3x-5$)

   c) -5, -7, -9 (equation is $y=-2x-1$)

   d) 16, 25, 36 (equation is $y=x^2$)

Level C

1. a) $A=\pi r^2$, non-linear

   b) $P=\frac{28}{3}t$, linear

   c) $C=\pi d$, linear

   d) $A=a^2$, non-linear

   e) $V=\frac{4}{3}\pi r^3$, non-linear

2. a) d: stopping distance

   s: speed of the car

   $d=ks^2$

   $32.5=k(60)^2$

$k=\frac{32.5}{60^2}$ $k=\frac{32.5}{60^2}=0.009$

$d=ks^2=0.009\times(120)^2=130$ m

b) $d=80$ $d=ks^2$

$s=\sqrt{\frac{d}{k}}=\sqrt{\frac{80}{0.009}}=94.14$ km/h

5.3 Answer Key

Level A

a) linear

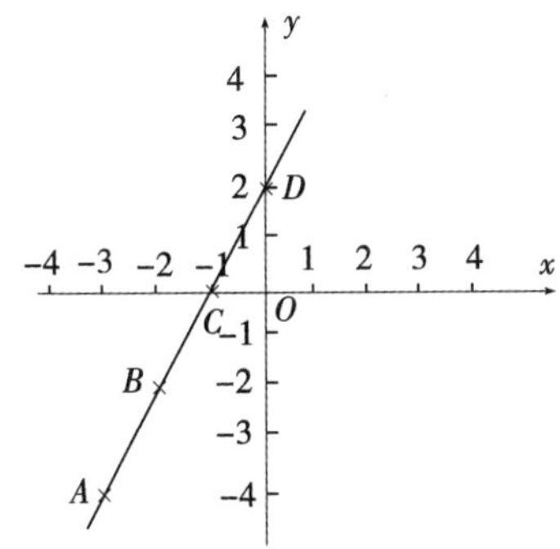

b) non-linear

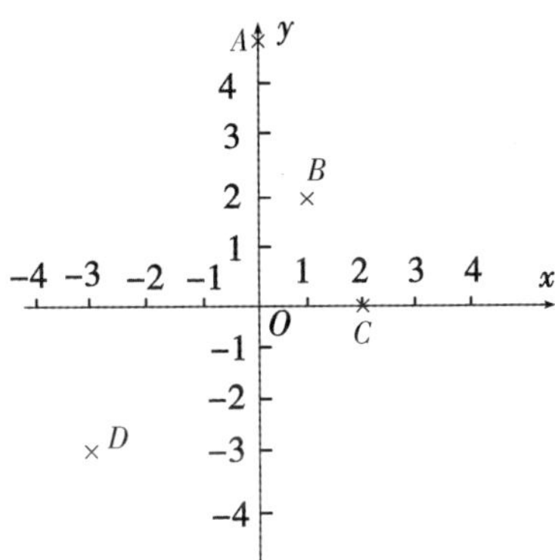

c) non-linear

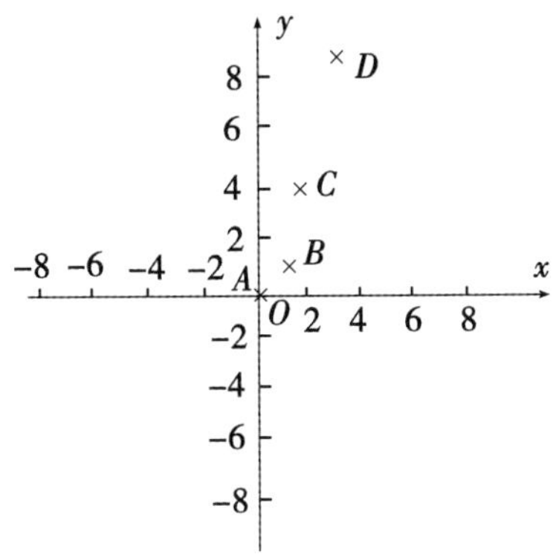

Level B

1. a) non-linear b) linear c) linear
   d) linear e) non-linear f) non-linear
   g) non-linear h) non-linear i) non-linear
   j) non-linear
2. a) $D(3,6)$ b) $D(1,-2)$
   c) $C(4,1)$ d) $A(-8,6)$
   e) $C(0,1)$ f) $D(1,10)$
   g) $B(5,-4)$ h) $D(2,6)$
   i) $D(6,4)$ j) $D(2,-9)$
3. a) 2 b) 3 c) 1 d) 3
   e) 3 f) 2 g) 6 h) −2
4. a) $y=x$ slope $=1$

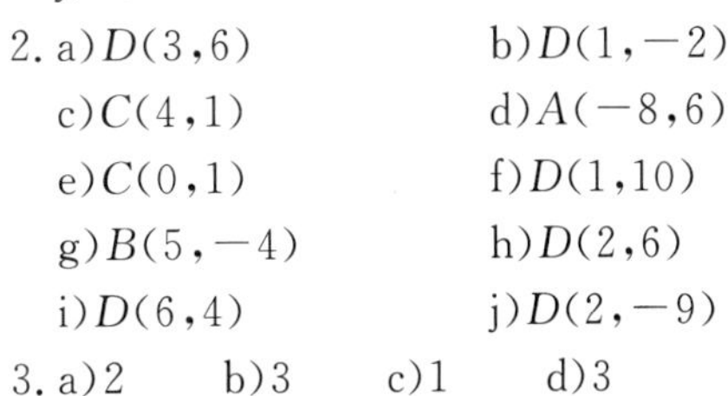

b) $y=x+2$ slope $=1$

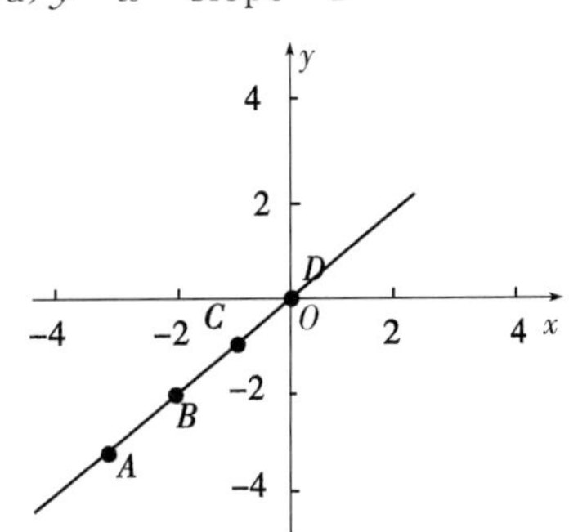

c) $y=\frac{1}{2}x+\frac{5}{2}$ slope $=\frac{1}{2}$

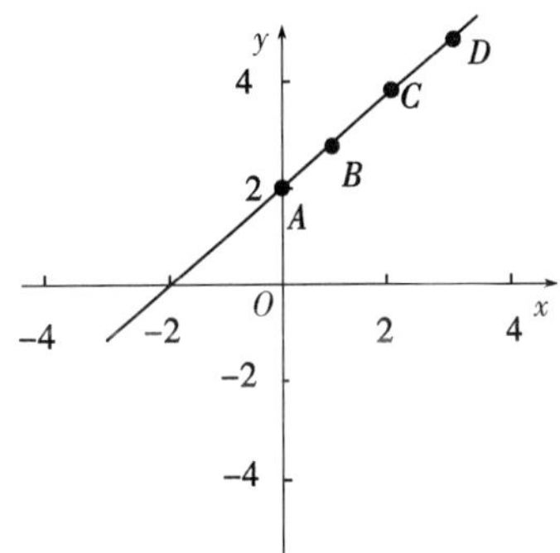

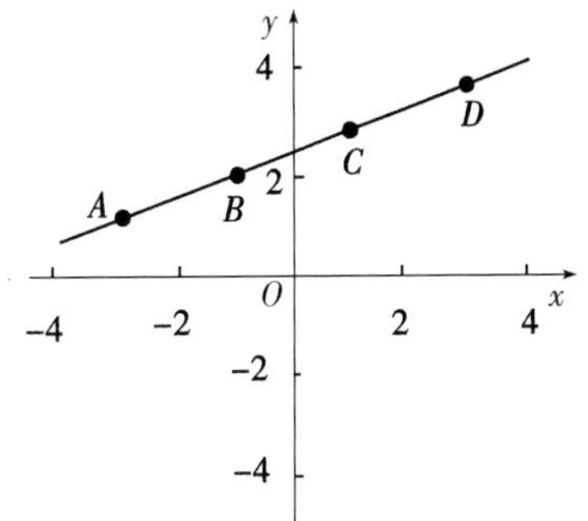

d) $y=-x+1$ slope $=-1$

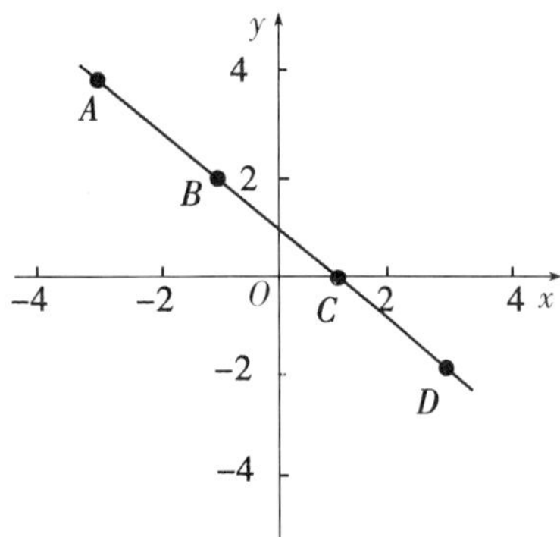

Level C

1. a) $\frac{245-140}{7-10}=\frac{105}{-3}=-35$

The rate of fuel consumption is 35 miles/gal.

b) x=Fuel remaining when started driving.

$\frac{140-0}{10-x}=-35$, $x=14$ gal

c) $y=mx+b$

$m=-35$, when $x=14$, $y=0$

$0=-35\times14+b$, $b=490$

$y=-35x+490$, or $D=-35F+490$

2. a) 300 m/min b) 31 min c) $D=300\ T$

Chapter 5 TEST

1. a) $-\frac{9}{2}$ b) 0 c) $-\frac{5}{12}$

d) $-\frac{13}{6}$ e) $\frac{14}{3}$ f) $\frac{5}{39}$

2. a) $y=2x$ linear

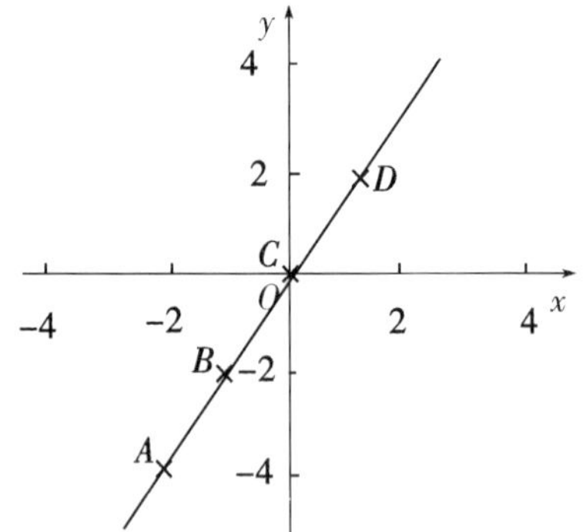

b) Not linear

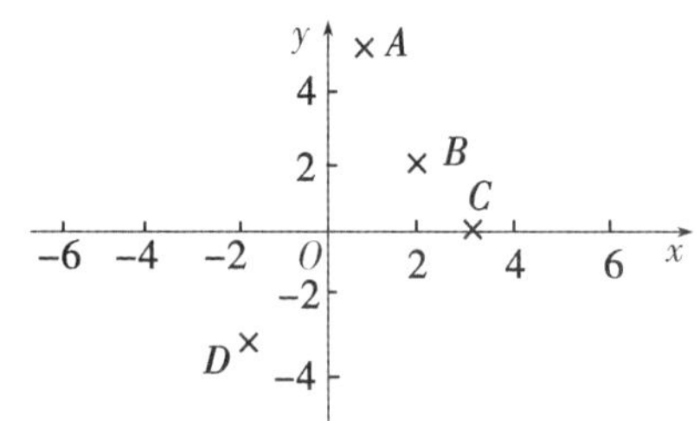

3. a)

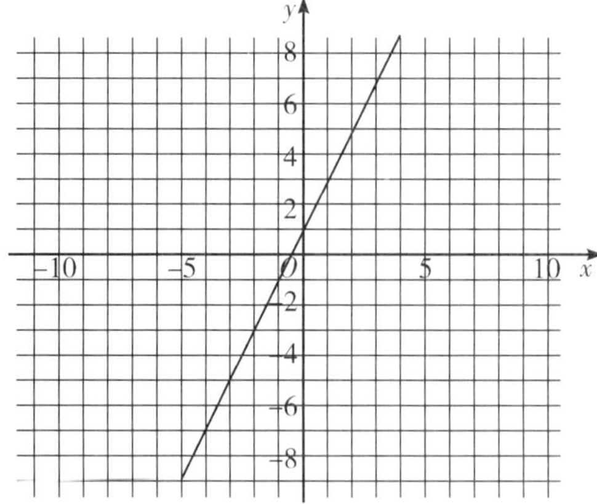

b)

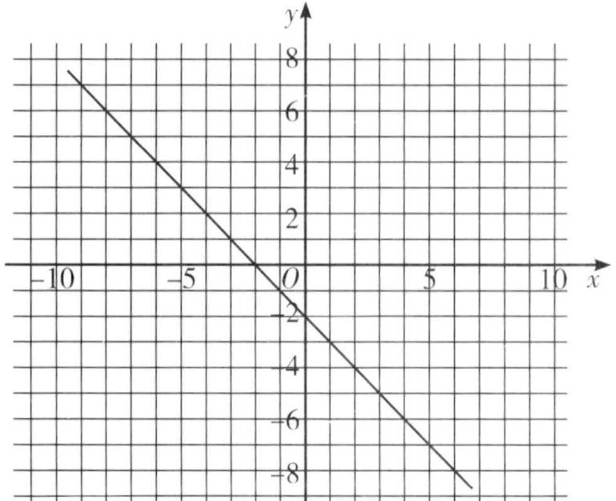

c)

y 8 6 4 2 -10 -5 0 5 10 x -2 -4 -6 -8

d)

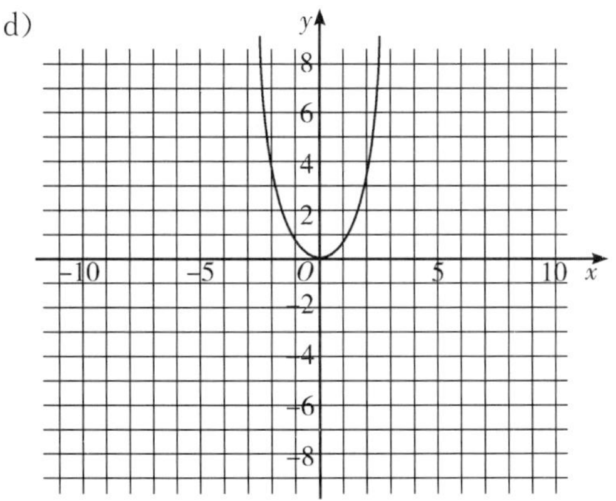

4. a) $-8, -10, -12$ (equation is $y=-2x-4$)

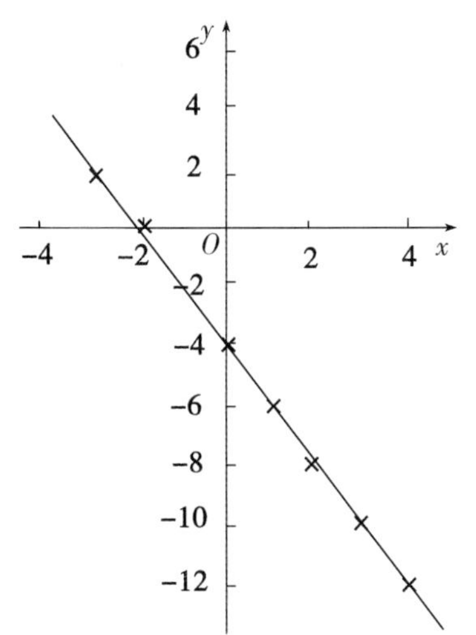

b) $-1, -2, -3$ (equation is $y=-\frac{1}{3}x+\frac{1}{3}$)

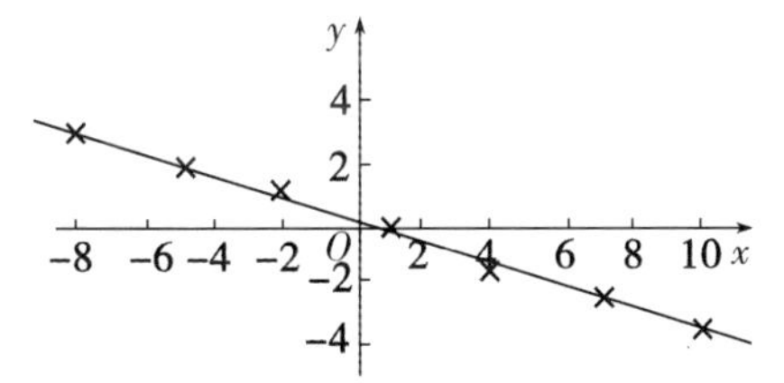

5. a)

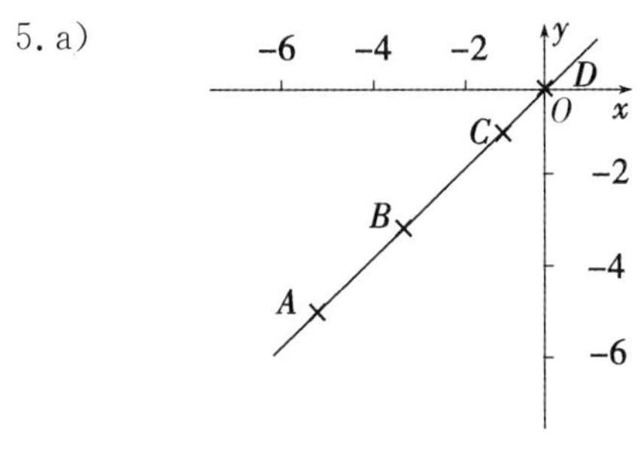

slope=1

b)

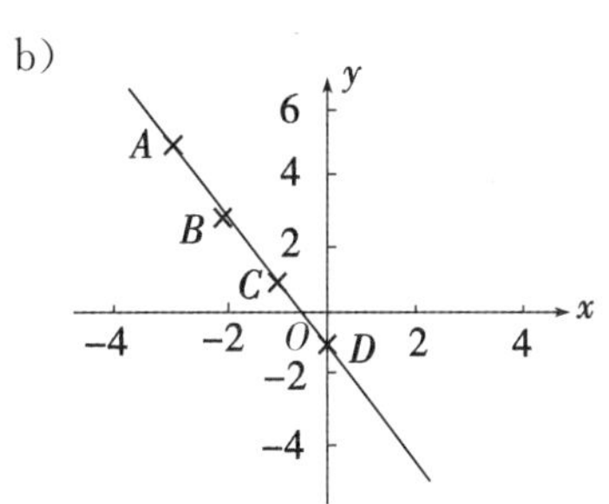

slope=2

c)

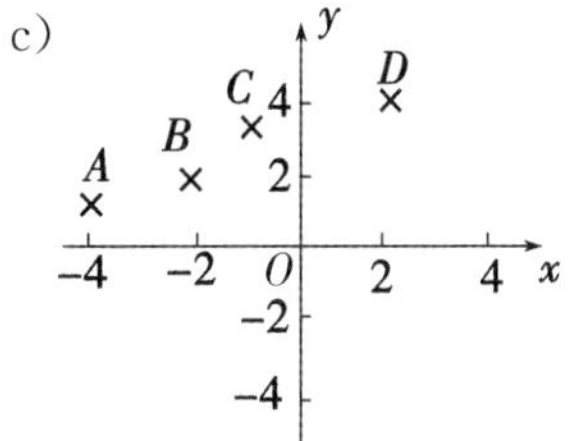

Not linear

d)

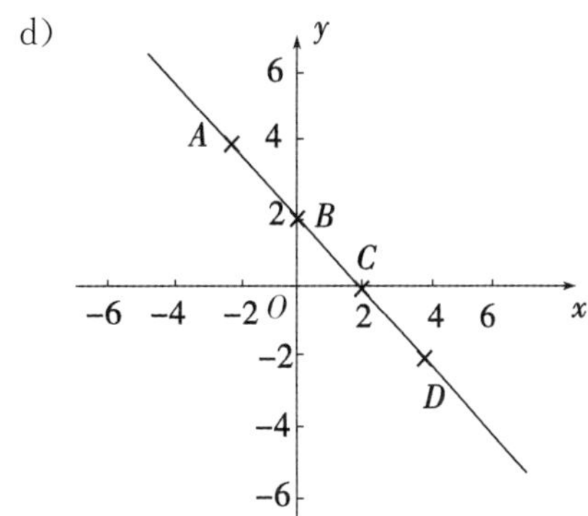

slope=-1

6. a) $113

b) 2143

ANSWERS CHAPTER 6

6.1 Answer Key

Level A

Number of Terms	Highest Degree of Terms	Coefficients	Like Terms	Constant Term
—	—	—	—	—
4	2	4 and 5 for xy, 3 for x	$4xy$ and $5xy$	-2
4	2	-1 and 2 for x^2, -1 and 3 for x	$-x^2$ and $2x^2$, $3x$ and $-x$	0
4	3	2 for a^2b, -2 for c^2, 3 for a^2, -2 for b	no	0
4	5	3 for x^2y^2, -6 and 3 for y^2, -1 for $uvxyz$	$-6y^2$ and $3y^2$	0
2	1	2.5×10^7 for a	no	2.1×10^8

Level B

1. a) $5xy-5x-4$ b) $5x+5y$
   c) $9xy+3x-2$ d) $7x-7y$
   e) $4x^2-4x$ f) $2x^2-8x+110$
   g) $6x^2y-6xy$ h) $8x^2y$
2. a) $-6xy+8x+5y-10$
   b) $10x^2+8xy$
   c) $-45x^2y-24xy+3x$
   d) $2x^2+4xy+2y^2$
   e) $-3x^2y-2xy-2x^2+3y$
   f) $2a^2b^2-2a^2b-8ab^2$
   g) $-6x^2y+12xy+8y$
   h) $-2a^2b-5ab^2$
   i) $2.3a\times10^7+2.22\times10^8$
   j) $3.2b\times10^3+2.2a\times10^3+2\times10^2$
3. a) $-48x^6y^7$ b) $12x^7y^7$
   c) $1296x^{10}y^{10}$ d) $-\frac{3}{2}\frac{x^4y^2}{z^4}$
   e) $\frac{1}{4}a^3b$ f) $\frac{2x^2z^4-x^2y^2}{4z^4}$
   g) $\frac{4a^2+2ab-6b^2}{9a^2+18ab+9b^2}$

Level C

a) $=4x^{2+\frac{2}{3}}y^{2+\frac{2}{5}}+\frac{4}{9}x^{1+\frac{5}{3}}y^{1+\frac{7}{5}}$

$=4x^{\frac{8}{3}}y^{\frac{12}{5}}+\frac{4}{9}x^{\frac{2}{3}}y^{\frac{12}{5}}$

$=\frac{40}{9}x^{\frac{8}{3}}y^{\frac{12}{5}}$

b) $=\frac{2}{5}\times\frac{4}{5}\cdot x^{\frac{2}{3}+\frac{4}{5}}y^{\frac{5}{2}+\frac{4}{3}}z^{\frac{3}{4}+\frac{2}{3}}+\frac{7}{10}\times\frac{1}{10}x^{\frac{1}{2}+\frac{1}{3}}$

$y^{\frac{1}{5}+\frac{3}{4}}z^{\frac{1}{4}+\frac{1}{3}}-\frac{2}{3}\times\frac{3}{4}x^{\frac{2}{5}+\frac{16}{15}}y^{\frac{1}{6}+\frac{11}{3}}z^{\frac{5}{4}+\frac{1}{6}}$

$=\frac{8}{25}x^{\frac{22}{15}}y^{\frac{23}{6}}z^{\frac{17}{12}}+\frac{7}{100}x^{\frac{5}{6}}y^{\frac{19}{20}}z^{\frac{7}{12}}-\frac{1}{2}x^{\frac{22}{15}}y^{\frac{23}{6}}z^{\frac{17}{12}}$

$=-\frac{9}{50}x^{\frac{22}{15}}y^{\frac{23}{6}}z^{\frac{17}{12}}+\frac{7}{100}x^{\frac{5}{6}}y^{\frac{19}{20}}z^{\frac{7}{12}}$

c) $\frac{y^{\frac{2}{3}}z^{\frac{7}{5}}}{x^{\frac{1}{2}}y^{\frac{1}{5}}z^{0.4}}\times\frac{x^{\frac{2}{9}}y^{\frac{3}{5}}z^{0.5}}{y^{0.8}z^{\frac{1}{4}}}$

$=\frac{x^{\frac{2}{9}}y^{\frac{2}{3}+\frac{3}{5}}z^{\frac{7}{5}+0.5}}{x^{\frac{1}{2}}y^{\frac{1}{5}+0.8}z^{0.4+\frac{1}{4}}}$

$=\frac{x^{\frac{2}{9}}y^{\frac{19}{15}}z^{\frac{19}{10}}}{x^{\frac{1}{2}}y^{1}z^{\frac{13}{20}}}$

$=x^{\frac{2}{9}-\frac{1}{2}}y^{\frac{19}{15}-1}z^{\frac{5}{4}}$

$=x^{-\frac{5}{18}}y^{\frac{4}{15}}z^{\frac{5}{4}}$

d) $\frac{\frac{13x^{\frac{3}{2}}y^{\frac{1}{3}}z^{0.1}}{51x^{0.3}y^{0.6}z^{0.8}}}{3x^{\frac{1}{5}}y^{\frac{1}{2}}z^{\frac{2}{3}}}$

$=13x^{\frac{3}{2}}y^{\frac{1}{3}}z^{0.1}\times\frac{3x^{\frac{1}{5}}y^{\frac{1}{2}}z^{\frac{2}{3}}}{51x^{0.3}y^{0.6}z^{0.8}}$

$=\frac{13}{17}x^{\frac{3}{2}+\frac{1}{5}-0.3}y^{\frac{1}{3}+\frac{1}{2}-0.6}z^{0.1+\frac{2}{3}-0.8}$

$=\frac{13}{17}x^{\frac{7}{5}}y^{\frac{7}{30}}z^{-\frac{1}{30}}$

6.2 Answer Key

Level A

1. a)3 b)3 c)1.2
   d)3 e)25 f)10
   g)5 h)0.3 i)27
   j)10 k)0.4 l)$1\frac{1}{9}$
   m)-36 n)-15 o)$\frac{2}{3}$
   p)$-\frac{1}{15}$

2. $\frac{620}{31}=20$ pages/day

Level B

1. a)1.6 b)1 c)-2
   d)10 e)$1\frac{23}{24}$ f)$18\frac{2}{3}$
   g)-5 h)19 i)10
   j)10 k)17 l)20
   m)3 n)-94 o)$\frac{2}{5}$

2. a)$\frac{1}{2}x+5=25, x=40$
   b)$\frac{1}{4}x+10=40, x=120$
   c)$75-(25+x)=200, x=-150$
   d)$3x+5=50, x=15$
   e)$120+\frac{x}{9}=125, x=45$

3. \$1.7/mL

Level C

a)$A=\frac{(a+b)h}{2}, 2A=(a+b)h$

$a+b=\frac{2A}{h}$

$a=\frac{2A}{h}-b$

b)$d=V_it+\frac{1}{2}at^2$

$\frac{1}{2}at^2=d-V_it$

$a=\frac{2(d-V_it)}{t^2}$

c)$\frac{1}{R_t}=\frac{1}{R_1}+\frac{1}{R_2}+\frac{1}{R_3}$

$\frac{1}{R_2}=\frac{1}{R_t}-\frac{1}{R_1}-\frac{1}{R_3}$

$=\frac{R_1R_3}{R_tR_1R_3}-\frac{R_tR_3}{R_tR_1R_3}-\frac{R_tR_1}{R_tR_1R_3}$

$=\frac{R_1R_3-R_tR_3-R_tR_1}{R_tR_1R_3}$

$R_2=\frac{R_tR_1R_3}{R_1R_3-R_tR_3-R_tR_1}$

d)$d_0=\frac{fd_i}{d_i-f}$

$d_0(d_i-f)=fd_i$

$d_0d_i-d_0f=fd_i$

$d_0d_i-fd_i=d_0f$

$(d_0-f)d_i=d_0f$

$d_i=\frac{d_0f}{d_0-f}$

e)$T=2\pi\sqrt{\frac{m}{k}}$

$T^2=4\pi^2\cdot\frac{m}{k}$

$m=\frac{T^2k}{4\pi^2}=k(\frac{T}{2\pi})^2$

6.3 Answer Key

Level A

1. a)15 b)5 c)-8
   d)-9 e)288 f)-20
   g)2 h)-1.8 i)$14\frac{2}{3}$
   j)$4\frac{8}{9}$ k)9.6 l)$-3\frac{3}{7}$
   m)-87 n)$7\frac{3}{11}$ o)0

2. 18 cm

Level B

1. a)-3 b)1.25 c)$\frac{1}{3}$
   d)2.5 e)-5.75 f)-2.5
   g)-40 h)$2\frac{2}{5}$ i)$9\frac{7}{9}$
   j)$4\frac{24}{41}$

2. 95 km/h

3. $4\frac{4}{9}$ hours

4. 26 km/h; 2 km/h

Level C

a) $0.5(\frac{x}{4}+\frac{2}{3})=0.2(\frac{x}{3}-\frac{3}{4})$

$0.5(\frac{x}{4}+\frac{2}{3})\times 12=0.2(\frac{x}{3}-\frac{3}{4})\times 12$

$0.5(3x+8)=0.2(4x-9)$

$1.5x+4=0.8x-1.8$

$0.7x=-4-1.8=-5.8$

$x=\frac{-5.8}{0.7}=\frac{58}{7}=-8\frac{2}{7}$

b) $\frac{x+2}{4}-\frac{x-1}{2}=\frac{2}{3}$

$12(\frac{x+2}{4}-\frac{x-1}{2})=\frac{2}{3}\times 12$

$3(x+2)-6(x-1)=8$

$3x+6-6x+6=8$

$-3x=-4$

$x=\frac{4}{3}$

c) $4+\frac{2}{x}=7+\frac{3}{x}$

$\frac{2}{x}-\frac{3}{x}=7-4$

$\frac{2-3}{x}=3$

$\frac{-1}{x}=3$

$-1=3x$

$x=-\frac{1}{3}$

d) $\frac{3}{2x}=\frac{4}{3x}-\frac{1}{2}$

$\frac{3}{2x}\times 6=(\frac{4}{3x}-\frac{1}{2})\times 6$

$\frac{9}{x}=\frac{8}{x}-3$

$\frac{9}{x}-\frac{8}{x}=-3$

$\frac{1}{x}=-3$

$x=-\frac{1}{3}$

Chapter 6 TEST

1. a) $4x^3y-4xy$

   b) $6x^2y$

   c) $9x^2-4x^2y+2y-3xy$

   d) $-2a^2b-8ab^2+2a^2b^2$

   e) $-6x^2y+9xy+15y$

   f) $-8a^2b-2a^2b^2-6ab^2$

2. a) $\frac{1}{2}$ b) -53 c) -15

   d) 10 e) 6 f) 30

   g) 12 h) $\frac{7}{2}$ i) -94

   j) -88 k) $7\frac{17}{69}$ l) 8

3. a) -4 b) $1\frac{15}{19}$ c) $\frac{1}{3}$

   d) 3 e) $-3\frac{2}{3}$ f) $\frac{3}{8}$

   g) $\frac{2}{3}$ h) $4\frac{64}{149}$ i) $3\frac{3}{31}$

   j) $4\frac{2}{61}$

4. a) 56 b) 160 c) 40

   d) 15 e) -425

5. 114 km/hr

6. $4\frac{4}{9}$ hours

7. ship = 30 km/hr; current = 2 km/hr

8. a) $x^{\frac{5}{12}}y^{\frac{23}{40}}z^{\frac{41}{70}}$

   b) $x^{3.4}y^{\frac{1}{30}}z^{-\frac{1}{30}}$

ANSWERS CHAPTER 7

7.1 Answer Key

Level A

1. a) 31.62 cm
   b) 48 cm
   c) 43.30 cm
   d) 37.08 cm
2. a) 25.90 cm
   b) 40.31 cm
   c) 109.09 cm
   d) 89.02 cm
3. height = 11.06 m
   perimeter = 57.27 m
   area = 132.66 m^2
4. $\sqrt{11-8x-3x^2}$

Level B

1. $4\frac{8}{13}$
2. 5
3. 18 m
4. 42.4 m
5. 0.8 m
6. 16 inches
7. 7.4 inches
8. 24 cm
9. 13.6 km
10. 75.6 km
11. 0.55 hrs = 33 minutes

Level C

1. 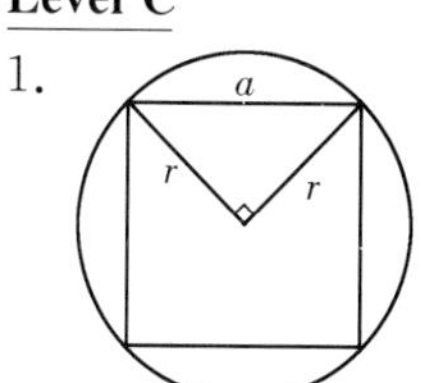

a: square side

$a^2=625, a=25$

$r^2+r^2=a^2=625$

$r^2=\frac{625}{2}$

circle area $=\pi r^2=\frac{625}{2}\pi$

2. 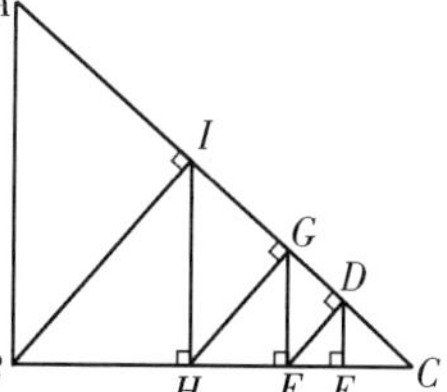

All triangles in the diagram are isosceles right triangles

$DE=EC=x$

$x^2+x^2=CD^2=2x^2$

$CD=DF$

$CD^2+DC^2=2CD^2=FC^2=4x^2$

$FC=FG$

$FC^2+FG^2=2FC^2=GC^2=8x^2$

$GC=GH$

$GC^2+GH^2=HC^2=2GC^2=16x^2$

$HC=IH, HC^2+IH^2=2HC^2=IC^2=32x^2$

$IC=IB, IC^2+IB^2=2IC^2=BC^2=64x^2$

$BC=AB, AB^2+BC^2=2BC^2=AC^2=128x^2$

$AC=4\sqrt{2}, AC^2=32$

$128x^2=32, x^2=\frac{32}{128}=\frac{1}{4}$

$x=\frac{1}{2}=0.5$

$DE=0.5$ cm

7.2 Answer Key

Level A

1. Perimeter = 119.39 cm
   Area = 667.74 cm^2
2. Perimeter = 61.42 cm
   Area = 235.62 cm^2
3. Perimeter = 141.70 cm
   Area = 1006.19 cm^2

Level B

1. 10.54 cm^2
2. $r=2$
3. 12.57 cm^2
4. 450 cm^2 & 1350 cm^2

5. 7 m^2

6. 12 cm

7. 15 m

8. 15 cm

9. 224 m^2

10. 20 cm

11. 20 cm

12. 48 cm

13. 12 cm

Level C

1. 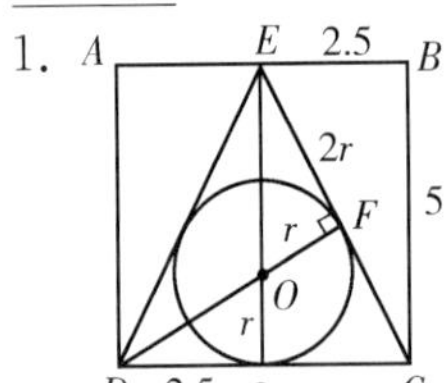

$\triangle EOF$ is similar to $\triangle ECB$

$\frac{EF}{FO}=\frac{BC}{EB}=\frac{5}{2.5}=\frac{2}{1}$

FO is radius. Assume $FO=r$.

$EF=2FO=2r$

$EO=\sqrt{FO^2+FE^2}=\sqrt{r^2+4r^2}=\sqrt{5}r$

$EO+OG=\sqrt{5}r+r=BC=5$

$(1+\sqrt{5})r=5$

$r=\frac{5}{\sqrt{5}+1}=1.55$

$DO=\sqrt{DG^2+OG^2}=\sqrt{2.5^2+r^2}=2.9$ cm

2. 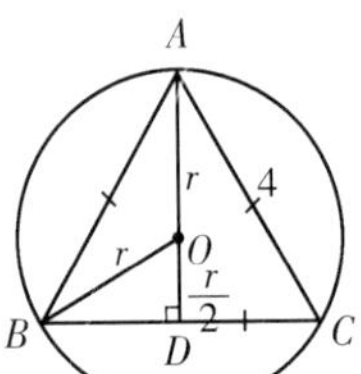

$OA=OB=r$(radius of the circle)

for $\triangle ABD$, $AB=4$, $BD=2$,

$AD=\sqrt{AB^2-BD^2}=2\sqrt{3}$

$\triangle ABD$ is similar to $\triangle BOD$

$\frac{AB}{AD}=\frac{BO}{BD}$, $\frac{4}{2\sqrt{3}}=\frac{r}{2}$

$r=\frac{4\times 2}{2\sqrt{3}}=\frac{4}{\sqrt{3}}=2.3$ cm

7.3 Answer Key

Level A

a) Pent $=540°$; Rect $=360°$; Tri $=180°$; Square $=360°$

b) Pent $=108°$; Rect $=90°$; Tri $=60°$; Square $=90°$

c) A Regular Pentagon cannot

d) 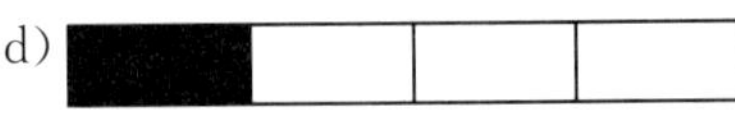

e) 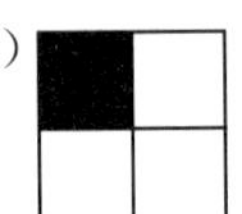

f) 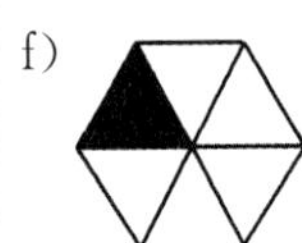

Level B

1. $(a-b)^2=\frac{c^2}{9}$

2. Length $=30$ cm
   Width $=10$ cm
   Area $=300$ cm^2

3. Answers Vary

4. A、C、D、E、F、G、H

5. Answers Vary

6. a)

b) 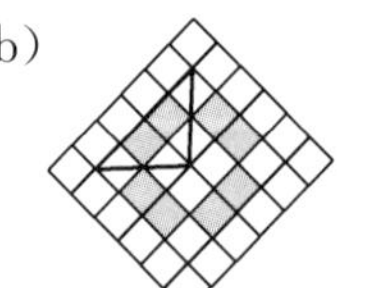

c) 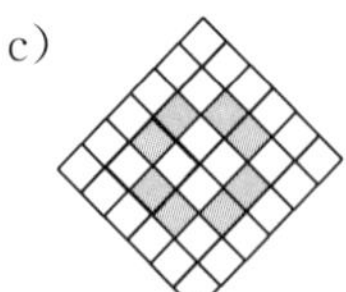

d) 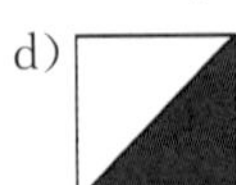

Level C

1. For a n-sided polygon,

   the sum of interior angles is

   $(n-2)\times180^\circ$ that is a multiple of 180°.

   So we just need to find the next number that is a multiple of 180° but greater than 1999°.

   $\frac{1999^\circ}{180^\circ}=11.1$

   So the sum of angles should be

   $180^\circ\times12=2160^\circ$

   The miss counting angle is $2160^\circ-1999^\circ=161^\circ$.

2.

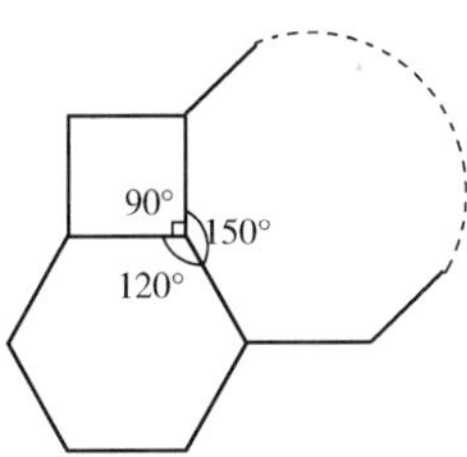

   From the diagram we know the interior angles of the third n-sided polygon must be 150°.

   $\frac{(n-2)\times180^\circ}{n}=150^\circ$

   $180^\circ(n-2)=150^\circ n$

   $180^\circ n-360^\circ=150^\circ n$

   $30^\circ n=360^\circ$

   $n=12$

   It's a regular dodecagon.

Chapter 7 TEST

1. Height = 24 cm

   Perimeter = 120 cm

   Area = $600\ cm^2$

2. $y=\sqrt{5x^2-36x+7}$

3. 12.99 cm

4. 10.625

5. Perimeter = 138.54 cm

   Area = $1178.10\ cm^2$

6. Perimeter = 196.00 cm

   Area = $2042.90\ cm^2$

7. $127.39\ m^2$

8. 2

9. $246.49\ cm^2$

10. $40.5\ cm^2$ & $121.5\ cm^2$

11. $140\ cm^2$

12. 4 cm

13. 23 cm

14. 17.78 cm

15. $864\ m^2$

16. 33.24 cm

17. a) F1

F2

F3

F4

b) Answers Vary

18.

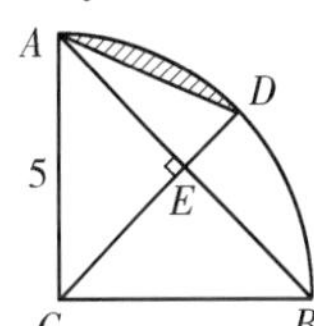

CD is a perpendicular bisector of AB, so $AE=EB$

$\angle ACE=45^\circ$, $\angle CAE=45^\circ$

$\therefore AE=EC$

$AE^2+EC^2=2EC^2=AC^2=5^2$

$EC=\frac{5\sqrt{2}}{2}$

the area of $\triangle ACD=\frac{CD\times AE}{2}=\frac{5\times\frac{5}{2}\sqrt{2}}{2}$

$=\frac{25\sqrt{2}}{4}$

the area of sector $ACD=\frac{1}{8}\pi\times5^2=\frac{40\pi}{8}$

the shaded area $=\frac{40\pi}{8}-\frac{25\sqrt{2}}{4}$

$=\frac{40\pi-50\sqrt{2}}{8}$ cm^2

19. 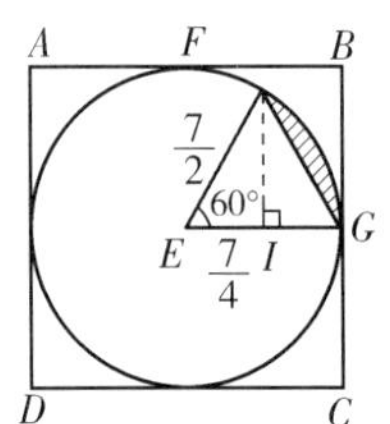

$HE=EG=\frac{7}{2}, EI=\frac{EH}{2}=\frac{7}{4}$

$HI=\sqrt{HE^2-EI^2}=\sqrt{(\frac{7}{2})^2-(\frac{7}{4})^2}$

$=\frac{7\sqrt{3}}{4}$

the area of $\triangle HEG$

$=\frac{EG\times HI}{2}=\frac{\frac{7}{2}\times\frac{7\sqrt{3}}{4}}{2}=\frac{49\sqrt{3}}{16}$

The area of sector HGE is $\frac{1}{6}$ of the area of the circle, so

$\frac{1}{6}\pi(\frac{7}{2})^2=\frac{49\pi}{24}$

the shaded area$=\frac{49\pi}{24}-\frac{49\sqrt{3}}{16}=1.1$ cm^2

20. 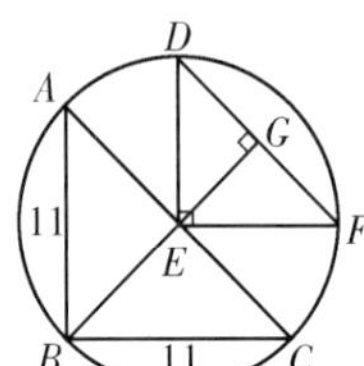

$AC=\sqrt{AB^2+BC^2}=\sqrt{11^2+11^2}=11\sqrt{2}$

radius$=\frac{AC}{2}=\frac{2\sqrt{11}}{2}=\frac{11}{2}\sqrt{2}$

$BE=DE=$radius$=\frac{11}{2}\sqrt{2}$

$\triangle ACB$ is similar to $\triangle DEG$

$\therefore \frac{AC}{BC}=\frac{DE}{EG}$,

$\frac{11\sqrt{2}}{11}=\frac{\frac{11\sqrt{2}}{2}}{EG}, EG=\frac{11}{2}$

$BG=BE+EG=\frac{11}{2}\sqrt{2}+\frac{11}{2}+\frac{11}{2}$

$=13.28$ cm

21. 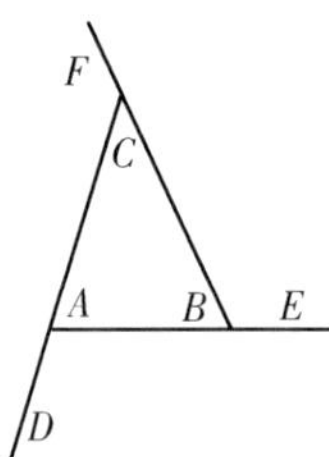

since the sum of exterior angles of any polygon is 360°

$\angle D=\frac{4}{4+5+6}\times 360=96°$

$\angle E=\frac{5}{4+5\times 6}\times 360=120°$

$\angle F=\frac{6}{4+5+6}=144°$

$\therefore \angle A=180°-96°=84°$

$\angle B=180°-120°=60°$

$\angle C=180°-144°=36°$

$\angle A:\angle B:\angle C=84:60:36=7:5:3$

ANSWERS CHAPTER 8

8.1 Answer Key

Level A

1. cylinder

2. 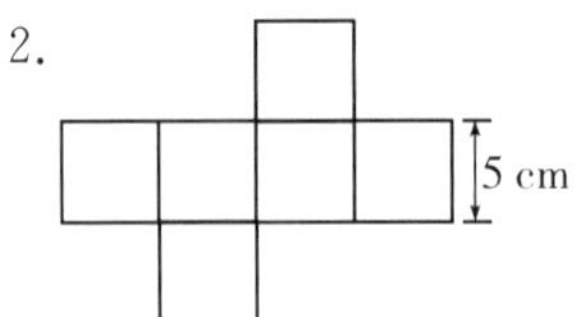

3. A、B、C、D

Level B

1. Pentagonal Pyramid

2. A. 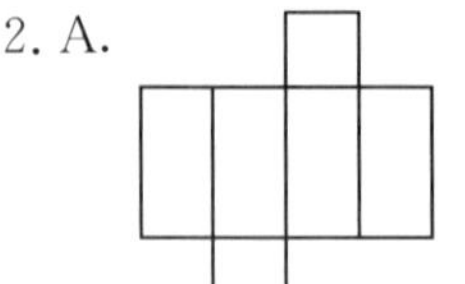

B.

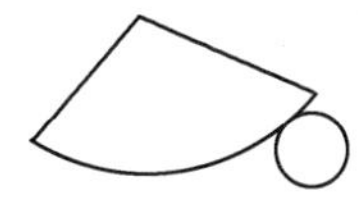

C.

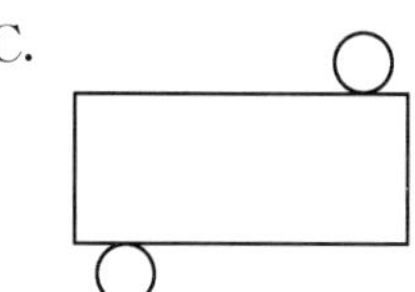

D.

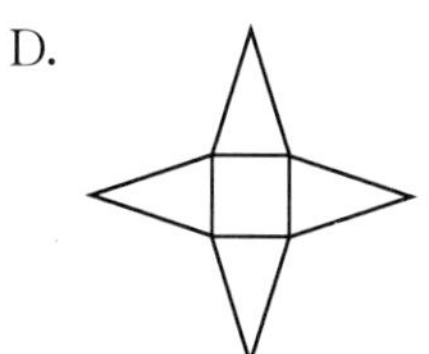

3.

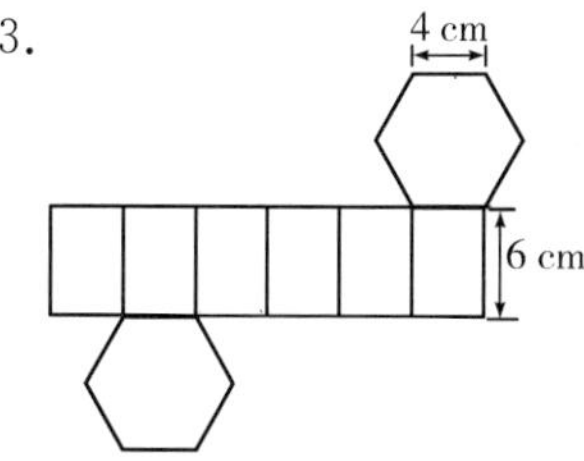

4.

5.

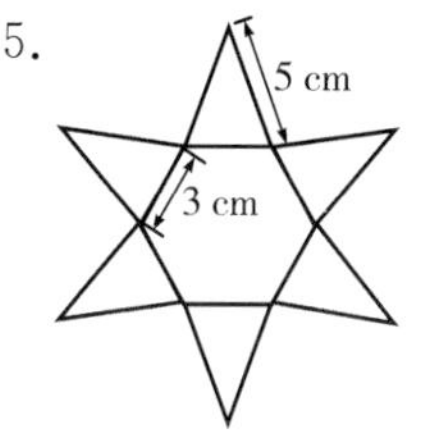

Level C

1.

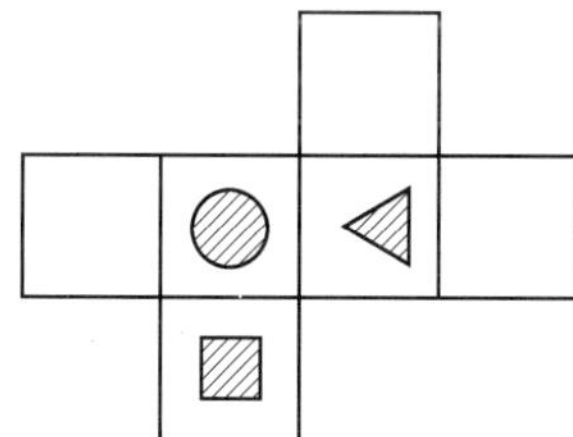

(Answers may vary)

2.

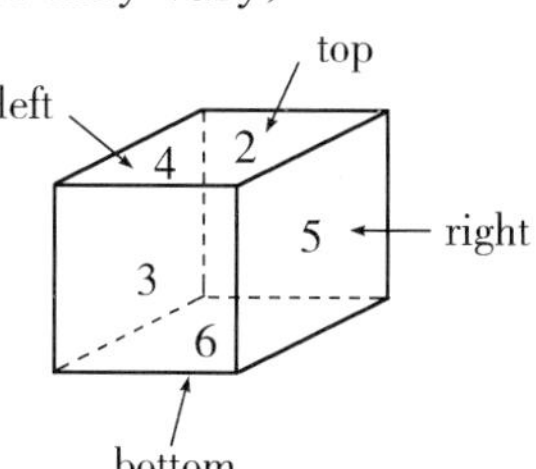

The greatest product is $3\times5\times6=90$.

3.

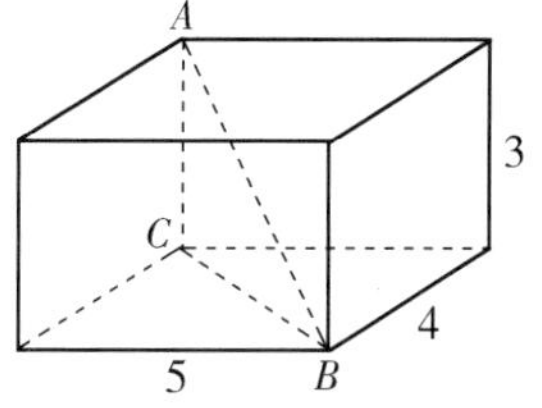

$AB^2=AC^2+BC^2$

$AC=3$

$BC^2=4^2+5^2$

$AB^2=3^2+4^2+5^2=50$

$AB=\sqrt{50}=5\sqrt{2}$

8.2 Answer Key

Level A

1. 13 600 cm^2
2. 10 m^2
3. 600π $cm^2=1885$ cm^2

Level B

1. 320π $cm^2=1005$ cm^2
2. $(4700+975\pi)$ $cm^2=7763$ cm^2
3. $2820
4. 6 cm
5. Width=1.5 m Height=5.22 m
6. $r=1.12$ cm $V=39.41$ cm^2
7. 18.65 m^2
8. 8.14 cm
9. 1862.04 cm^2

Level C

1. $2\pi R^2+2\pi R\times4+4\times3\times3$

$=2\pi\times8^2+2\pi\times8\times4+36$

$=128\pi+64\pi+36$

$=192\pi+36$

$=639.2$ cm^2

2. $2\pi(\frac{10}{2})^2-2\times3^2+4\times3\times5+2\pi(\frac{10}{2})\times5$

$=50\pi-18+60+50\pi$

$=356.2$ cm^2

8.3 Answer Key

Level A

1. 37.7 m^3
2. 1200 boxes
3. 8 m

4. $3000\pi\ cm^3 = 9425\ cm^3$

Level B

1. $170\ 000\ cm^3 = 0.17\ m^3$
2. $47\ 641.46\ cm^3$
3. 5 m
4. Width = 5.5 cm Height = 6 cm
5. $r = 3.16$ cm $SA = 261.16\ cm^2$
6. $82\ 331.85\ cm^3$
7. $h = 13.99$ cm $SA = 2732.52\ cm^2$
8. $V = 768\ 000\ cm^3$

Level C

1. $\pi R^2 \times 4 + 3 \times 3 \times 3$

$= \pi \cdot 8^2 \times 4 + 27$

$= 831.2\ cm^3$

2. $\pi(\frac{10}{2})^2 \times 5 - 3 \times 3 \times 5$

$= 125\pi - 45$

$= 347.7\ cm^3$

3.

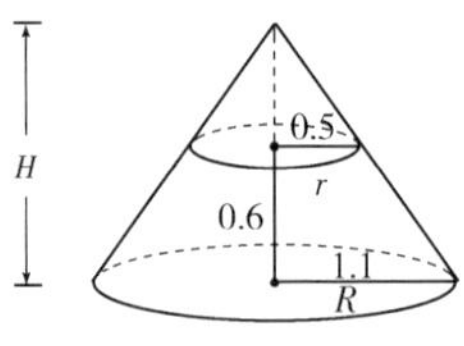

h, H, 0.5, 0.6, 1.1

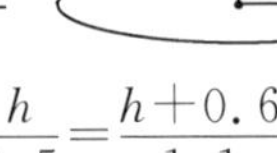

$\frac{h}{0.5} = \frac{h+0.6}{1.1}$

$1.1h = 0.5(h+0.6)$

$= 0.5h + 0.3$

$0.6h = 0.3$

$h = 0.5, H = 0.5 + 0.6 = 1.1$

Volume = Large Cone − Small Cone + Cylinder

$= \frac{1}{3}\pi R^2 H - \frac{1}{3}\pi r^2 h + \pi R^2 \times 1.5$

$= \frac{1}{3}\pi 1.1^3 - \frac{1}{3}\pi 0.5^3 + \pi 1.1^2 \times 1.5$

$= 6.96\ m^3$

Chapter 8 TEST

1. 8 m
2. 8 m
3. 2500 cases
4. 16 m
5. $S = 4000\pi\ cm^2$ $V = 34\ 375\pi\ cm^3$
6. a)

1 m, 1.6 m

b) $14.796\ m^2$

c) $4.16\ m^3$

7. a)

b) $4640.24\ cm^2$

c) $20\ 943.95\ cm^3$

8. $V = 0.17\ m^3$ $SA = 1.98\ m^2$
9. $V = 97\ 699.11\ cm^3$ $SA = 12\ 426.55\ cm^2$
10. 2.1 m
11. Width = 17 cm Height = 12 cm
12. $r = 11$ cm $SA = 2141.48\ cm^2$
13. $V = 174\ 314.15\ cm^3$ $SA = 22\ 231.42\ cm^2$
14. $h = 99.95$ cm $SA = 6160.94\ cm^2$
15. $V = 1.47\ m^3$ $SA = 10.1\ m^2$
16. $V = 1\ 093\ 274.24\ cm^3$ $SA = 57\ 992.68\ cm^2$
17. a) $256.31 cm^2$ b) $259.81 cm^3$
18. a) T b) T c) F d) T
19.

F E F A D E F G B C H G G H

20. a) c)
21. 60°
22. 74π
23. c)

ANSWERS CHAPTER 9

9.1 Answer Key

Level A

1.

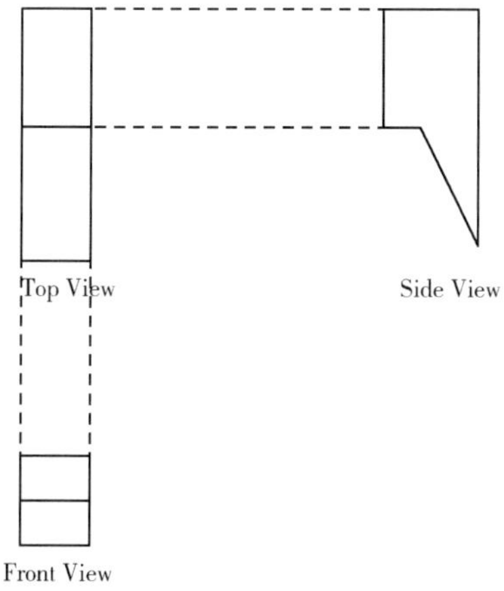

Top View

Side View

Front View

2.

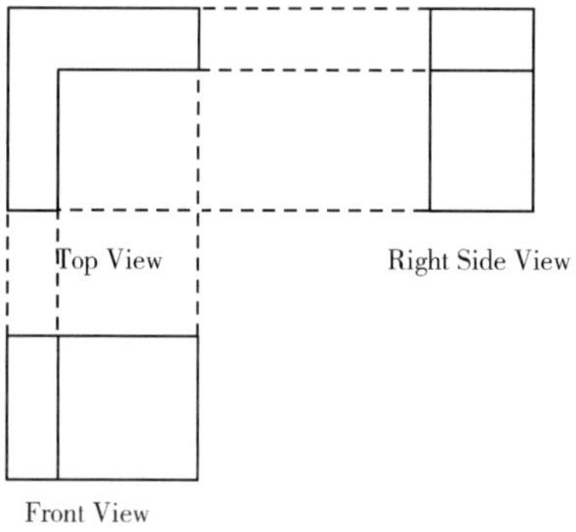

Top View

Right Side View

Front View

Level B

1.

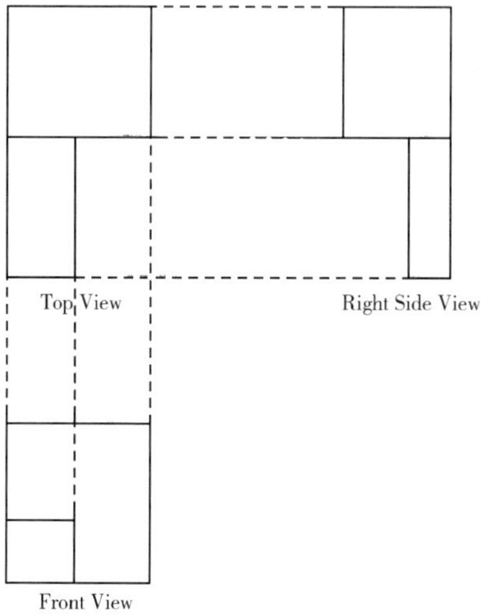

Top View

Right Side View

Front View

2.

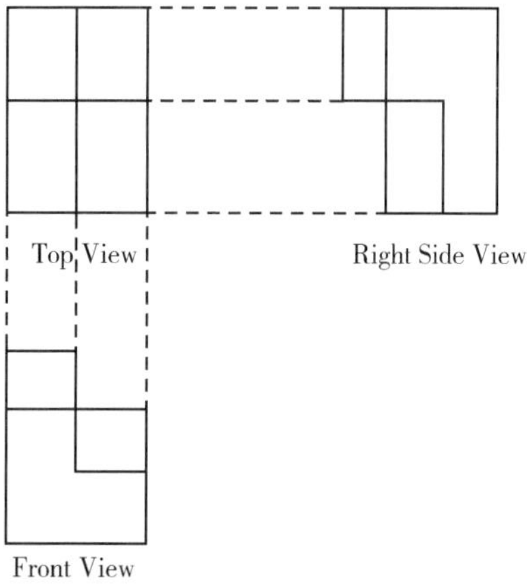

Top View

Right Side View

Front View

3.

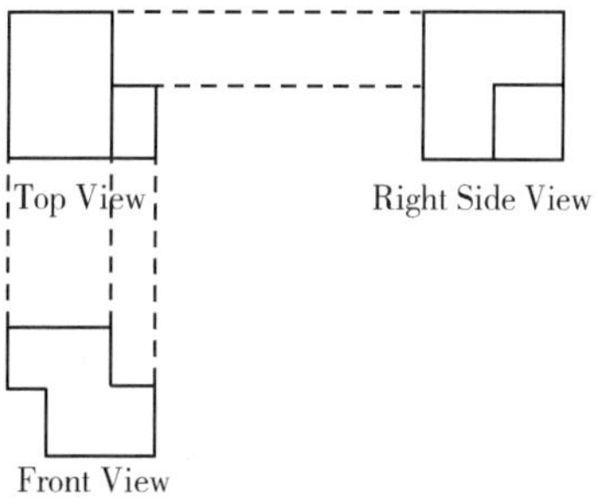

Top View

Right Side View

Front View

Level C

1.

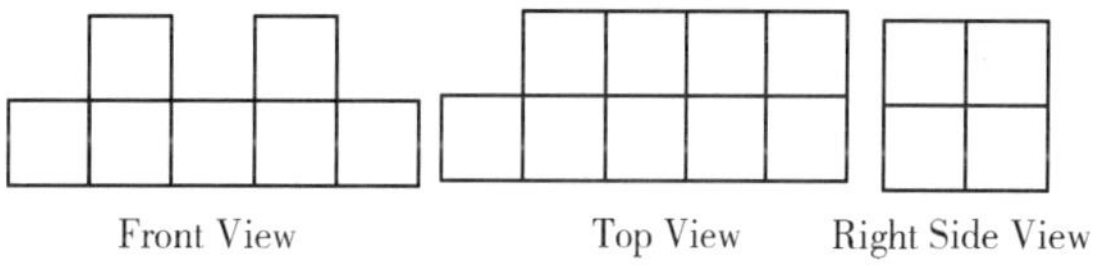

Front View

Top View

Right Side View

2.

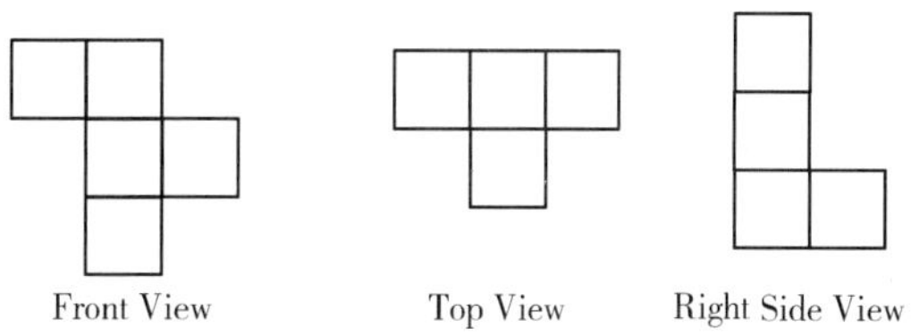

Front View

Top View

Right Side View

9.2 Answer Key

Level A

1. B

2.

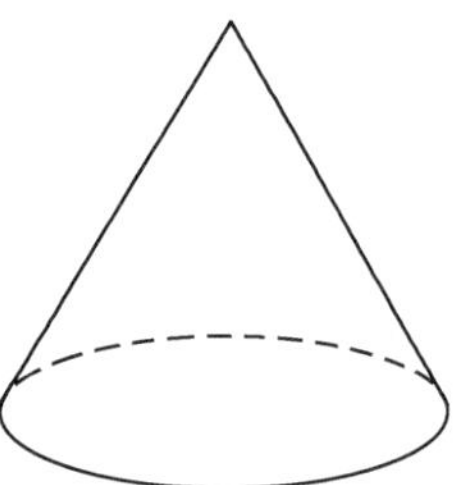

3.

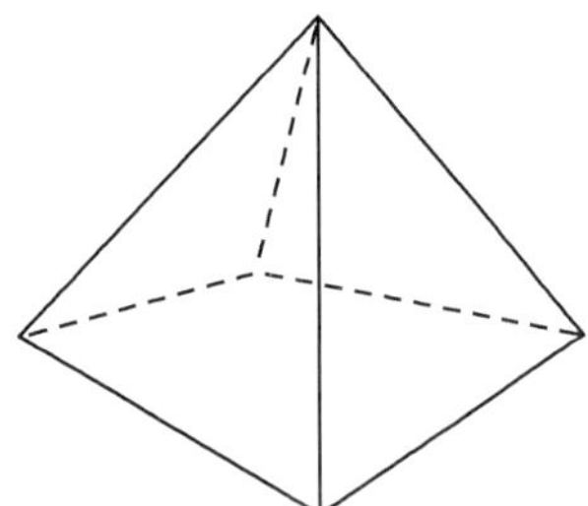

Level B

1.

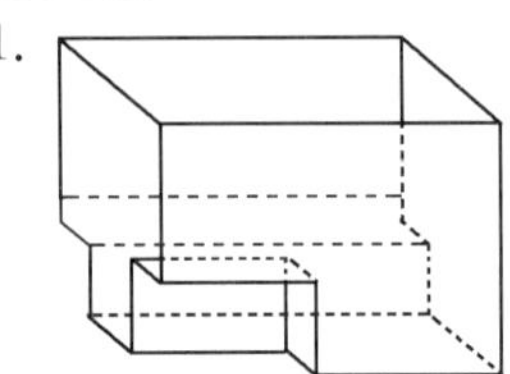

2.

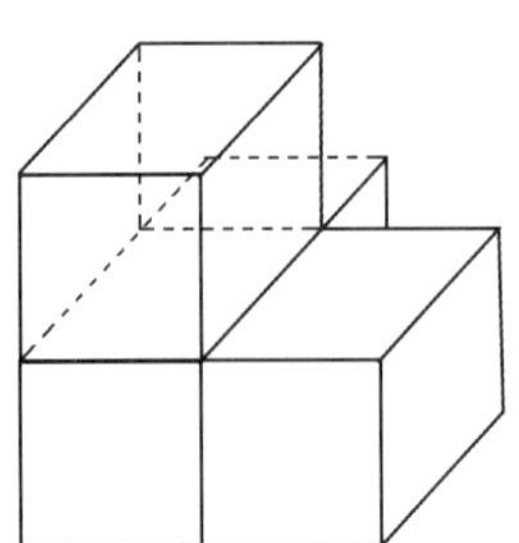

3.

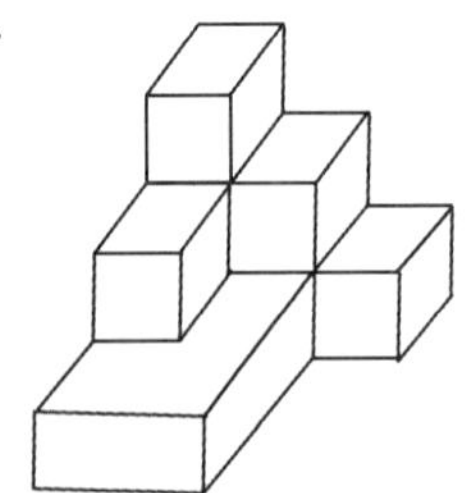

4.

Level C

1.

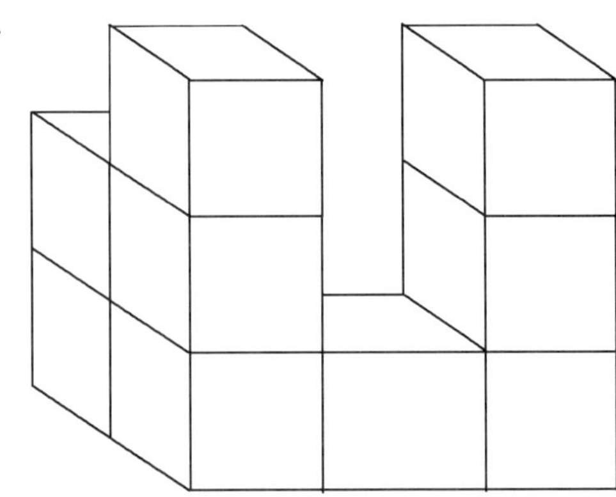

2.

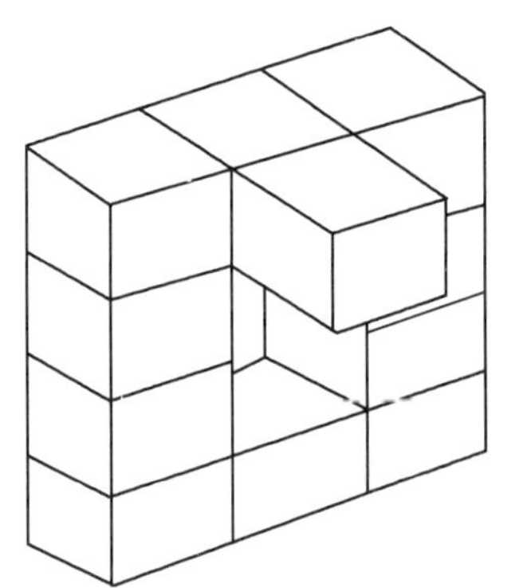

9.3 Answer Key

Level A

1.

Front View

Right Side View

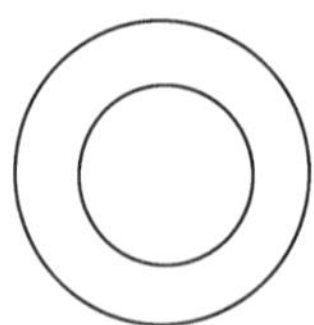

Top View

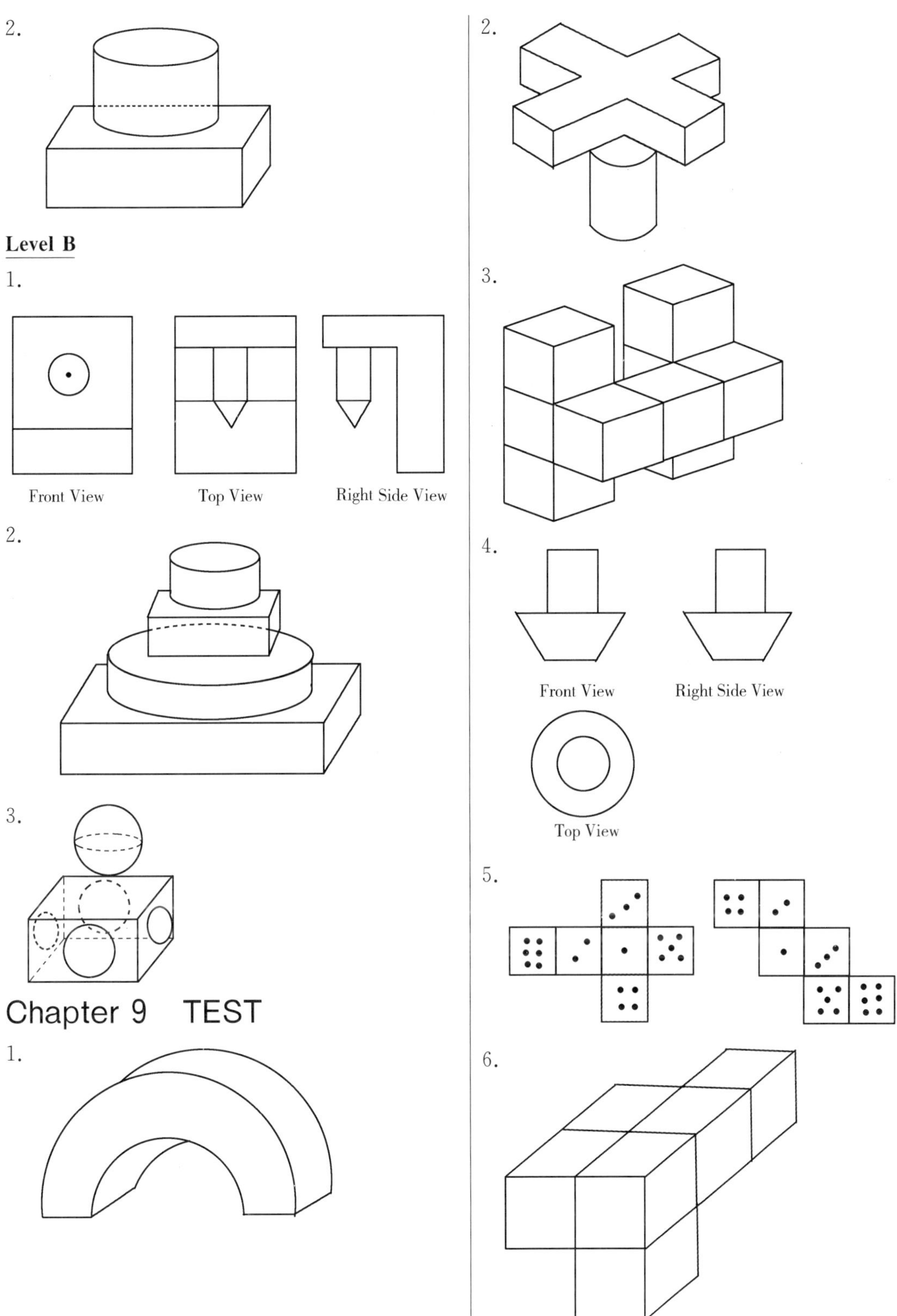
2.
Level B
1.
Front View
Top View
Right Side View
2.
3.
Chapter 9 TEST
1.
2.
3.
4.
Front View
Right Side View
Top View
5.
6.

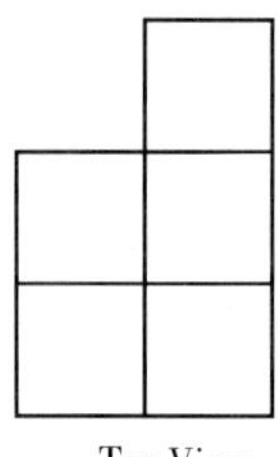

Top View

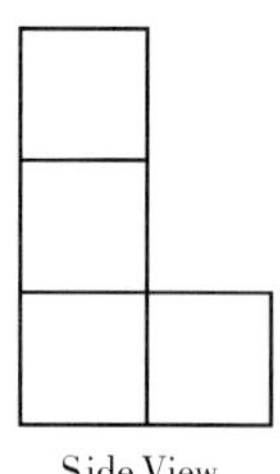

Side View

7.

Top View

8. b

ANSWERS CHAPTER 10

10.1 Answer Key

Level A

1. a) $\frac{1}{2}$ b) $\frac{2}{3}$ c) $\frac{1}{2}$

 d) 1 e) $\frac{2}{3}$ f) $\frac{1}{3}$

2. $\frac{5}{750}$ or $\frac{1}{150}$, $0.00\overline{6}$, $0.\overline{6}\%$

3. a) $\frac{15}{36}=\frac{5}{12}$ b) $\frac{13}{36}$

 c) $\frac{8}{36}=\frac{2}{9}$ d) 0

 e) $\frac{23}{36}$ f) $\frac{28}{36}=\frac{7}{9}$

 g) $\frac{23}{36}$ h) $\frac{21}{36}=\frac{7}{12}$

4. a) $\frac{12}{26}=\frac{6}{13}$ b) $\frac{14}{26}=\frac{7}{13}$

5. a) $\frac{2}{5}$ b) $\frac{3}{6}=\frac{1}{2}$

 c) $\frac{4}{11}$ d) $\frac{5}{9}$

 e) $\frac{3}{11}$ f) $\frac{5}{11}$

Level B

1. a)

Second / First	1	2	3	4	5	6
1	(1,1)	(1,2)	(1,3)	(1,4)	(1,5)	(1,6)
2	(2,1)	(2,2)	(2,3)	(2,4)	(2,5)	(2,6)
3	(3,1)	(3,2)	(3,3)	(3,4)	(3,5)	(3,6)
4	(4,1)	(4,2)	(4,3)	(4,4)	(4,5)	(4,6)
5	(5,1)	(5,2)	(5,3)	(5,4)	(5,5)	(5,6)
6	(6,1)	(6,2)	(6,3)	(6,4)	(6,5)	(6,6)

b)

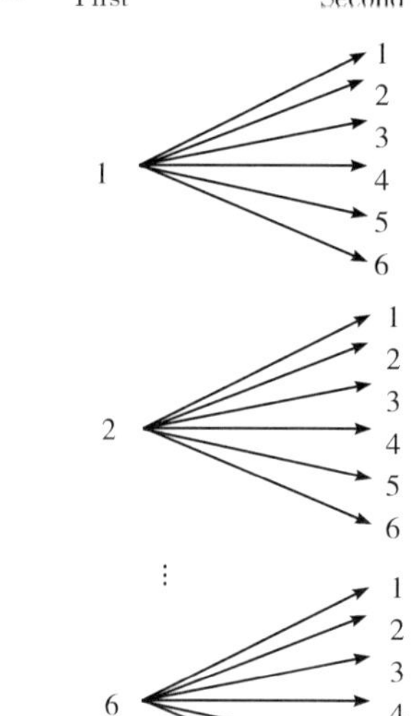

⋮

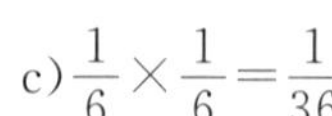

c) $\frac{1}{6}\times\frac{1}{6}=\frac{1}{36}$ d) $\frac{1}{2}\times\frac{1}{2}=\frac{1}{4}$

e) $\frac{1}{2}\times\frac{1}{2}=\frac{1}{4}$ f) $\frac{1}{2}\times\frac{1}{2}=\frac{1}{4}$

g) $\frac{1}{6}\times\frac{1}{6}=\frac{1}{36}$ h) $\frac{20}{36}=\frac{5}{9}$

i) $\frac{4}{36}=\frac{1}{9}$

2. a) $\frac{1}{2}$ b) $\frac{6}{52}=\frac{3}{26}$

c) $\frac{1}{2}$

d) $\frac{40}{52}=\frac{10}{13}$

e) $\frac{8}{52}=\frac{2}{13}$

f) $\frac{22}{52}=\frac{11}{26}$

Level C

1. The tree diagram of the first three putting is:

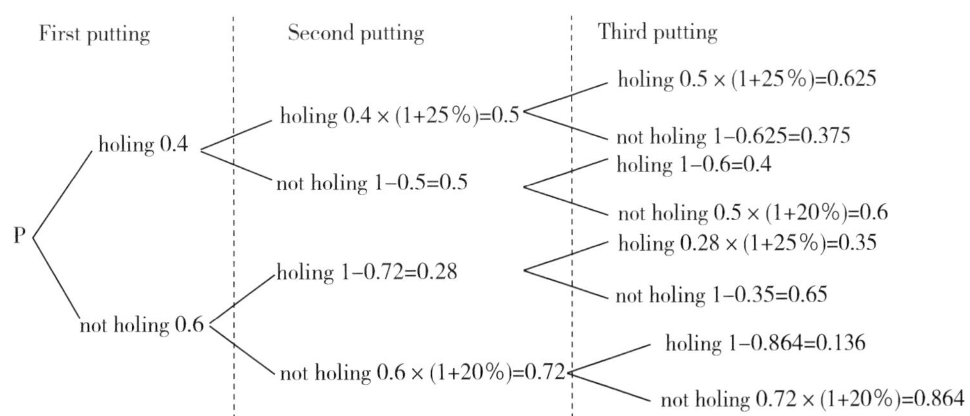

a) P = P(first putt holing) × P(second putt holing)
$=0.4\times0.5=0.2$

b) P = P(first not holing) × P(second not holing)
$=0.6\times0.72=0.432$

c) $P=0.4\times0.5\times0.375+0.4\times0.5\times0.4+0.6\times0.28\times0.35$
$=0.2138$

2. assume $P(D_i)=0.6$ is the probability of being caught in a traffic jam in Day-i(D_i).

$P(\bar{D}_i)=1-P(D_i)=0.4$

P(at least two out of three day)

= P(two out of three days) or P(three out of three days)

$=P(D_1)P(D_2)P(\bar{D}_3)+P(D_1)P(\bar{D}_2)P(D_3)+P(\bar{D}_1)P(D_2)P(D_3)+P(D_1)P(D_2)P(D_3)$

$=0.4\times0.6\times0.6\times3+0.6\times0.6\times0.6$

$=0.648$

10.2 Answer Key

Level A

1. a) $\frac{1}{8}$ b) $\frac{1}{8}$

c) $\frac{3}{8}$ d) $\frac{3}{8}$

e) $\frac{7}{8}$ f) $\frac{1}{2}$

2. a) $\frac{4}{36}=\frac{1}{9}$ b) $\frac{1}{36}$

c) $\frac{10}{36}=\frac{5}{18}$ d) $\frac{15}{36}=\frac{5}{12}$

e) $\frac{6}{36}=\frac{1}{6}$ f) $\frac{15}{36}=\frac{5}{12}$

g) $\frac{9}{36}=\frac{1}{4}$ h) 0

Level B

1. a) $\frac{1}{9}$ b) $\frac{16}{81}$

c) $\frac{16}{81}$ d) $\frac{8}{81}$

e) $\frac{4}{9}$ f) $\frac{49}{81}$

g) $\frac{16}{81}$ h) $\frac{32}{81}$

i) $\frac{5}{9}$ j) $\frac{4}{27}$

2. $\frac{2}{n+1}$

3. a) $\frac{1}{12}$ b) $\frac{1}{6}$

c) $\frac{2}{9}$ d) $\frac{1}{9}$

e) $\frac{5}{12}$ f) $\frac{7}{12}$

g) $\frac{1}{6}$ h) $\frac{5}{12}$

i) $\frac{7}{12}$ j) $\frac{1}{6}$

Level C

1. The Venn diagram of the problem is:

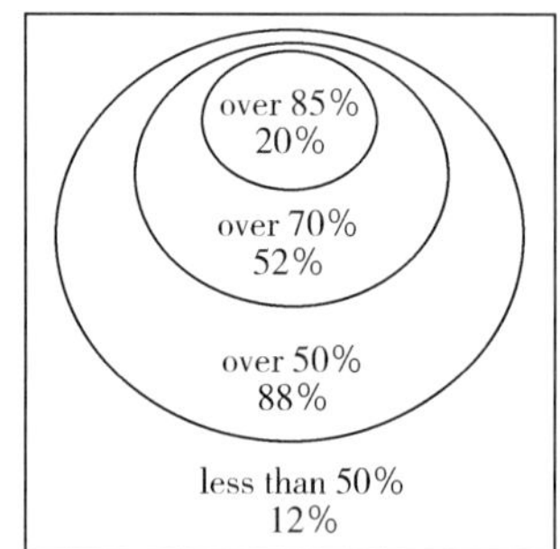

so 88% of students earned 50% or above

2. The Venn diagram of the problem is:

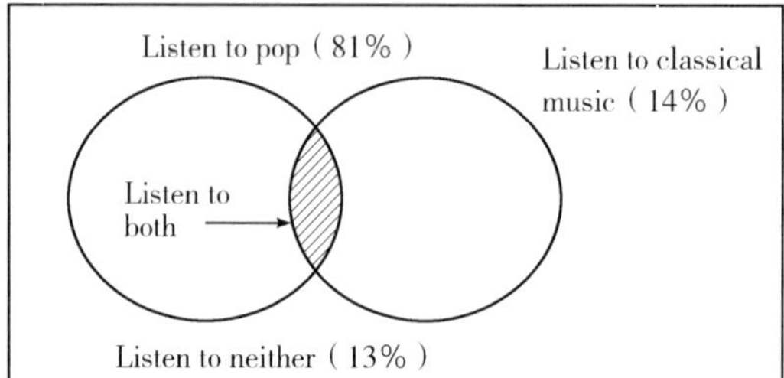

$81\%+14\%-(1-13\%)$

$=95\%-87\%=8\%$

Chapter 10 TEST

1. a) $\frac{65}{216}$ b) $\frac{65}{432}$

c) $\frac{4}{81}$ d) $\frac{169}{1296}$

e) $\frac{529}{1296}$ f) $\frac{65}{216}$

g) $\frac{25}{144}$ h) $\frac{5}{27}$

2. a) $\frac{13}{42}$ b) $\frac{13}{84}$

c) $\frac{2}{45}$ d) $\frac{13}{105}$

e) $\frac{253}{630}$ f) $\frac{13}{42}$

g) $\frac{1}{6}$ h) $\frac{4}{21}$

3. a) $\frac{1}{9}$ b) $\frac{1}{36}$

c) $\frac{1}{4}$ d) 1

e) $\frac{3}{4}$ f) $\frac{1}{2}$

g) $\frac{5}{12}$ h) $\frac{1}{3}$

i) $\frac{5}{6}$ j) $\frac{4}{9}$

4. a) $\frac{1}{4}$ b) $\frac{3}{26}$

c) $\frac{1}{2}$ d) $\frac{10}{13}$

e) $\frac{2}{13}$ f) $\frac{5}{26}$

5. a) $\frac{1}{2}$ b) $\frac{1}{4}$

c) $\frac{1}{4}$ d) $\frac{100}{169}$

e) $\frac{9}{169}$ f) $\frac{1}{16}$

g) $\frac{1}{169}$ h) $\frac{100}{169}$

6. a) $\frac{26}{51}$ b) $\frac{13}{51}$

c) $\frac{25}{102}$ d) $\frac{10}{17}$

e) $\frac{11}{221}$ f) $\frac{13}{204}$

g) $\frac{1}{221}$ h) $\frac{95}{663}$

7. $\frac{2}{15}$

8. $\frac{2}{17}$

Excellence in Math Workbook Series

Order Form 2013–2014

Title	ISBN	Unit Price (C$)	Copies	Subtotal(C$)
Mathematics 8 Workbook	978-7-5537-1843-9	32.95		
Mathematics 9 Workbook	978-7-5537-1844-6	32.95		
Pre-Calculus 10 Workbook	978-7-5537-1845-3	36.95		
Pre-Calculus 11 Workbook	978-7-5537-1846-0	36.95		
Pre-Calculus 12 Workbook	978-7-5537-1847-7	36.95		
		Plus 5% GST		
		Total (C$)		

Please contact 604.721.9698 or grace@doyenpublishing.ca for a quote of shipping and handling charges.

Please make cheque or money order payable to Doyen Publishing Inc.

Doyen Publishing Inc.

865 W. 70th Avenue, Vancouver, BC V6P 2X6

Tel: 604.721.9698

Fax: 604.323.9698

Email: grace@doyenpublishing.ca

Website: www.doyenpublishing.ca

图书在版编目(CIP)数据

数学提优作业 = Excellence math workbook. 八年级:英文/张诚编. --南京:江苏科学技术出版社,2013.8
ISBN 978-7-5537-1843-9

Ⅰ.①数… Ⅱ.①张… Ⅲ.①中学数学课-初中-习题集-英文 Ⅳ.①G634.605

中国版本图书馆 CIP 数据核字(2013)第198938号

Excellence Math Workbook Grade 8

主　编	Chris Zhang, Ph. D.
责任编辑	熊亦丰
助理编辑	卢海春
责任校对	郝慧华
责任监制	曹叶平　周雅婷
出版发行	凤凰出版传媒股份有限公司 江苏科学技术出版社
出版社地址	南京市湖南路1号A楼,邮编:210009
出版社网址	http://www.pspress.cn
经　销	凤凰出版传媒股份有限公司
印　刷	安徽省新华印刷有限公司
开　本	880mm×1230mm　1/16
印　张	19
字　数	456 000
版　次	2013年10月第1版
印　次	2013年10月第1次印刷
标准书号	978-7-5537-1843-9
定　价	197.70元

图书如有印装质量问题,可随时向我社出版科调换。